T0243048

CAMBRIDGE LIBRARY COLLECTION

Books of enduring scholarly value

History of Medicine

It is sobering to realise that as recently as the year in which On the Origin of Species was published, learned opinion was that diseases such as typhus and cholera were spread by a 'miasma', and suggestions that doctors should wash their hands before examining patients were greeted with mockery by the profession. The Cambridge Library Collection reissues milestone publications in the history of Western medicine as well as studies of other medical traditions. Its coverage ranges from Galen on anatomical procedures to Florence Nightingale's common-sense advice to nurses, and includes early research into genetics and mental health, colonial reports on tropical diseases, documents on public health and military medicine, and publications on spa culture and medicinal plants.

The Life of Edward Jenner M.D.

Written by his friend, the physician John Baron (1786–1851), this laudatory biography of the 'father of immunology' did much to enhance the reputation of Edward Jenner (1749–1823) upon its publication in two volumes between 1827 and 1838. The work covers Jenner's personal and professional life both before and after his development of the vaccine for smallpox, as well as touching on the vaccine's reception and use around the world. Thoroughly explaining the history and facts of vaccination, Baron established himself as an authority on the subject. Although criticised by some for its unquestioning praise of Jenner's genius, the work is valuable for its use of primary sources, drawing heavily on correspondence and personal notes, excerpts of which appear throughout the text. Volume 2, published in 1838, covers Jenner's later life and the global reception of vaccination. The appendix lists the various honours bestowed upon him.

Cambridge University Press has long been a pioneer in the reissuing of out-of-print titles from its own backlist, producing digital reprints of books that are still sought after by scholars and students but could not be reprinted economically using traditional technology. The Cambridge Library Collection extends this activity to a wider range of books which are still of importance to researchers and professionals, either for the source material they contain, or as landmarks in the history of their academic discipline.

Drawing from the world-renowned collections in the Cambridge University Library and other partner libraries, and guided by the advice of experts in each subject area, Cambridge University Press is using state-of-the-art scanning machines in its own Printing House to capture the content of each book selected for inclusion. The files are processed to give a consistently clear, crisp image, and the books finished to the high quality standard for which the Press is recognised around the world. The latest print-on-demand technology ensures that the books will remain available indefinitely, and that orders for single or multiple copies can quickly be supplied.

The Cambridge Library Collection brings back to life books of enduring scholarly value (including out-of-copyright works originally issued by other publishers) across a wide range of disciplines in the humanities and social sciences and in science and technology.

The Life of Edward Jenner M.D.

With Illustrations of his Doctrines,
and Selections from his Correspondence

VOLUME 2

JOHN BARON

CAMBRIDGE
UNIVERSITY PRESS

CAMBRIDGE
UNIVERSITY PRESS

University Printing House, Cambridge, CB2 8BS, United Kingdom

Published in the United States of America by Cambridge University Press, New York

Cambridge University Press is part of the University of Cambridge.
It furthers the University's mission by disseminating knowledge in the pursuit of
education, learning and research at the highest international levels of excellence.

www.cambridge.org
Information on this title: www.cambridge.org/9781108071147

This edition first published 1838
This digitally printed version 2014

ISBN 978-1-108-07114-7 Paperback

Sir T Lawrence, P. R. A.   W. H. Mote.

Edw. Jenner

M.D. - F.R.S.

FISHER, SON & Cᵒ LONDON & PARIS, 1838.

THE

LIFE

OF

EDWARD JENNER, M.D.

LL. D., F. R. S.

PHYSICIAN EXTRAORDINARY TO HIS MAJESTY GEO. IV.

FOREIGN ASSOCIATE OF THE NATIONAL INSTITUTE OF FRANCE,

&c. &c. &c.

WITH

ILLUSTRATIONS OF HIS DOCTRINES,

AND

SELECTIONS FROM HIS CORRESPONDENCE.

BY

JOHN BARON, M.D., F.R.S.

LATE SENIOR PHYSICIAN TO THE GENERAL INFIRMARY, AND CONSULTING
PHYSICIAN TO THE LUNATIC ASYLUM AT GLOUCESTER, FELLOW OF THE
ROYAL MEDICAL AND CHIRURGICAL SOCIETY OF LONDON, &c. &c.

IN TWO VOLUMES.

VOL. II.

LONDON:

HENRY COLBURN, PUBLISHER,

GREAT MARLBOROUGH STREET.

1838.

J. B. NICHOLS AND SON, 25, PARLIAMENT STREET.

ADVERTISEMENT.

THOSE who have done me the favour of perusing the former Volume, are aware of the inducements which prompted me to publish it before the completion of the work. The part now brought forth finishes my undertaking. It has been in manuscript for several years. The reasons which have retarded the publication are so much of a private and personal nature as to render it unnecessary to specify them. This delay has been irksome to myself; but I hope it has not been injurious, either to the cause of vaccination, or to the character of its author.

The feelings avowed in the first volume have never been absent from my mind during any part of the progress of the work. I am deeply conscious of its many imperfections, and should have been well contented to have seen the whole subject handled by some one possessed of more leisure and greater powers than belong to me. The labour has been carried on amid many hindrances; and had not the kind individuals whose names I have mentioned in the introduction, given me their aid, I must have abandoned it entirely. I have great pleasure in again recording their good offices, especially those of Richard Gamble, my oldest friend, whose attachment to me has been " closer than that of a brother :" SED NEC ILLA EXTINCTA SUNT, ALUNTURQUE POTIUS ET AUGENTUR COGITATIONE ET MEMORIA.

He laboured with me in the examination of the voluminous papers with a devotion to the name of Jenner that well deserves to be had in remembrance : but, above all, he enabled me to give those illustrations of the views of pestilence which I deemed it necessary to present in the Fifth and Sixth Chapters of the first volume. Every succeeding day seems to have confirmed the truth of these views; and it was a notion, perhaps an unfounded one, that the dissemination of them would tend to augment the confidence in vaccination, which tempted me to commit them to the press before the whole work was ready. In these conclusions I may have erred, but the sincerity of my purpose will not, on that account, I trust, be called in question.

CONTENTS.

CHAPTER I.

CHAPTER II.

CHAPTER III.

CHAPTER IV.

CHAPTER V.

CHAPTER VI.

CHAPTER VII.

CHAPTER VIII.

CHAPTER IX.

ERRATA.

VOL. I.

Page   63, line 12 from bottom, *for* principal, *l.* principle.

Page 239, line 2 from top, *for* Gunning, *l.* Dunning.

Page 376, line 7 from top, *for* delineates, *l.* delineate.

Page 429, *l.* Dr. De Carro to Dr. Jenner, *for* to Dr. De Carro, Vienna.

Page 606, line 6 from top, *for* successfully, *l.* successively.

VOL. II.

Pages 10, 52, 53, *for* Frank, *l.* Franck.

Page   56, line 2 from top, *for* he, *l.* she.

Page 106, line 2 from top, *for* Valentine, *l.* Valentin.

Page 108, line last, *for* astonished, *l.* enraged.

Page 168, line 19 from top, dele *to*.

LIFE

OF

DR. JENNER.

CHAPTER I.

CONSEQUENCES OF THE FIRST PARLIAMENTARY GRANT—
JENNER RESIDENT IN LONDON—HIS ANTICIPATIONS
DISAPPOINTED — PROGRESS OF VACCINATION ABROAD,
AND OF OPPOSITION AT HOME—THE NAPOLEON MEDAL,
AND OTHER HONOURS — HIS ALLEGED DISTRUST OF
VACCINATION IN THE CASE OF HIS OWN SON.

THE discussion in parliament, and the very in-
adequate grant which was the result of it, by no
means produced the effects that Jenner's friends
anticipated. It stirred up greater hostility and
envy, and materially added to his own responsibility, without giving him the strength and inde-
pendence which might better have enabled him to
cope with his antagonists. He was left with the

whole weight of a most momentous undertaking
upon his own shoulders. Those who were jealous
of his fame waxed more bold; his friends became
lukewarm; his enemies more united and cla-
morous; the demands upon his time and attention
were increased; his private resources were diminished; and he could not devote himself to his
practice as a physician. Crippled and distressed
though he was by the very means which some
fondly imagined would have proved most beneficial to him, he, nevertheless, took his station and
kept it firmly. He fixed his mind upon the great
object which he was called upon to fulfil, and resolved at all hazards to persevere, and never to
desert the cause while he had power to labour in
it. In this attitude we shall ever find him. Had
he been more selfish, more ambitious, more desirous of pursuing objects of personal emolument or
aggrandizement, he certainly had the fairest opportunities of doing so; and no one could justly
say that any distinction, which such a man might
have acquired, was unmerited.

The people of England seemed to think that the
fee-simple both of his body and mind had been
purchased by the TEN THOUSAND POUNDS; and
many an unjust and ungenerous intimation of this
feeling was conveyed to him. To a mind like his,
this was no small annoyance. He was called upon
for explanations, for opinions, by every person who
thought a direct communication with the author

of Vaccination an honour worth seeking; when
they might have obtained all the information they
wanted from his published writings.

It is likewise to be remembered that a more
formidable and rancorous resistance than had yet
appeared, began to show itself; and had he not
been constantly cheered and animated by the conviction that the knowledge of his discovery was
rapidly extending itself over the earth, and that the
unceasing opposition of his enemies could not
interfere with the real and substantial benefits
which it was actually conferring, he would have
had many reasons to regret the conspicuous elevation on which it had placed him.

Influenced by the remarks of some of his parliamentary advocates, he was induced to fix himself
in Hertford-street, May Fair. The result of this
plan by no means corresponded with their anticipations. "Elated and allured," he observes, "by
the speech of the Chancellor of the Exchequer, I
took a house in London for ten years, at a high
rent, and furnished it; but my first year's practice
convinced me of my own temerity and impru-
dence, and the falsity of the minister's prediction.
My fees fell off both in number and value; for,
extraordinary to tell, some of those families in
which I had been before employed, now sent to
their own domestic surgeons or apothecaries
to inoculate their children, alleging that they
could not think of troubling Dr. Jenner about a

thing executed so easily as vaccine inoculation. Others, who gave me such fees as I thought myself entitled to at the first inoculation, reduced them at the second, and sank them still lower at the third." The truth is, that Jenner, in publishing his discovery as he did, effectually prevented the fulfilment of Mr. Addington's predictions; and it was scarcely befitting the representatives of a great nation to speculate on a contingency of this nature, in calculating the reward due to such a benefactor. He himself remarks to one of his correspondents, " I have now completely made up my mind respecting London. I have done with it, and have again commenced village-doctor. I found my purse not equal to the sinking of a thousand pounds annually (which has actually been the case for several successive years,) nor the gratitude of the public deserving such a sacrifice. How hard, after what I have done, the toils I have gone through, and the anxieties I have endured in obtaining for the world a greater gift than man ever bestowed on them before (excuse this burst of egotism), to be thrown by with a bare remuneration of my expenses!" *

* That some estimate may be formed of the nature of the treatment which he received, I subjoin the following extracts from letters written by him to an intimate friend.

June 3, 1804. " The Treasury still withholds the payment of what was voted me two years ago; and now there

Independently of these causes of distress, his mind was much agitated by anxiety respecting the health of Mrs. Jenner. She had been seized with spitting of blood, and this occurrence was the source of painful solicitude to him during the remainder of her life. It was deemed by his friends in London very desirable that he should be in town during the course of this spring (1804), to attend the anniversary of the Royal Jennerian Society, celebrated on his birth-day—the 17th of May. In declining a pressing invitation on this subject, he observes, " though a post-chaise or a mail-coach might bring up my body, my mind would be left behind. One cause of my absence, among many others, is the sad state of Mrs. Jenner's health. I cannot leave her even for a day with any comfort to my feelings. My friends, who honour the glorious cause of vaccination by assembling on the 17th, will, I trust, admit my apology. It is my intention to collect a few

are new officers, the time may be very long before a guinea reaches me from that quarter."

Nov. 2, 1804. " The London smoke, I have observed, is too apt to cloud our best faculties. I don't intend to risk the injury of mine in this way; except it may be occasionally, merely for the transaction of business. That the public has not the smallest right to expect it of me, no one will deny.— I have received no reward for showing them how to remove one of the greatest obstacles to human happiness; but, on the contrary, am loaded with a tax of more than £400 a year ! "

staunch vaccinists on that day at my cottage. I shall give them some roast beef, not forgetting a horn or two of good October. We shall close the day with bumpers of milk-punch to the health of the friends of humanity at the Crown and Anchor; and if it were not for the indisposition of my poor wife, we should roar like bulls."

The facts above recorded relative to remuneration, induced many of Dr. Jenner's friends to turn their eyes to other portions of the globe, which were benefiting largely by the vaccine discovery; with the hope that they would testify their gratitude by some substantial token. Our rich possessions in the East were first looked to on this occasion. The ever-active and benevolent Dr. Lettsom started the idea, and wrote to Jenner on the subject. His reply drew forth the following expression from another friend, the late Benjamin Travers, esq.

London, Feb. 18*th*, 1804.

My dear Doctor,

I have just read your interesting letter of the 8th inst. to Dr. Lettsom. It does you the greatest honour; and I shall make the best use of it that lies in my power. I wish I had had the happiness of your acquaintance a few years ago: you should not have acted in the manner you have: your liberality and disinterestedness every one must admire and extol; but you are sadly deficient in worldly wisdom.

I shall do the little which lies in my power to impress the minds of those whose influence may prove serviceable, was it merely to save the character of the *nation* from being blasted with ingratitude to a man to whom the WORLD has been so greatly indebted.

I am, my dear Doctor,

with the most unfeigned esteem and regard,

yours truly,

BENJAMIN TRAVERS.

In another letter, the same judicious correspondent observes, " If you had undertaken the extinction of the small-pox yourself, with coadjutors of your own appointment, I am confident you might have put £100,000 in your pocket; and the glory be as great, and the benefit to the community the same."

This excellent individual was not more mindful of Jenner's private affairs than he was of vaccination itself. As a member of the Jennerian Society, he was endeavouring to effect an object, which (one would have imagined) might have been accomplished without difficulty; I mean the abandonment of variolous inoculation at the Small Pox Hospital. But, strange to say, this act of justice and of mercy was delayed till nearly twenty years after the period of which I now write.

The Directors of the Vaccine Board, soon after this period, felt themselves called upon, in consequence of the peculiar situation of Dr. Jenner, to deviate somewhat from the ostensible purposes of their appointment, and to take some charge of his private concerns, as well as of the subject of vaccination. Sufficient evidence has been already given that the "amor sceleratus habendi" did not influence his actions; but, free as he himself was from all taint of this kind, his friends could not bear the reflection that his disinterestedness should actually lead to his personal loss at the very time that he was the instrument of conveying unheard-of benefits to mankind. A committee was appointed, in consequence of a reference from the Board of Directors, to enquire whether Dr. Jenner was not a sufferer in his income and pecuniary circumstances, "in consequence of the time which he had devoted to his valuable discovery of vaccine inoculation, and of the various expenses incident thereto, notwithstanding the parliamentary grant of £10,000," This investigation could not but be interesting to Dr. Jenner himself, and will of course justify me in dwelling upon it for a short time. The detail which I am about to subjoin, certainly does not afford any great encouragement to scientific men to divulge the result of their labours; but we have now this one great consolation, that in the exact proportion of the neglect which Jenner experienced from his contemporaries, did the purity, and firmness, and generosity of his principles shew themselves.

The account between him and the public stood thus :—Without entering into minute calculations,

or referring to other countries, it may be stated
that Dr. Jenner made known a discovery, which
has already materially increased the mean duration
of human life, and which was capable of rescuing
annually between thirty and forty thousand of
our own population from a pestilential and fatal
disease. To put the world fully in possession
of these blessings he abandoned almost entirely
the emoluments of his profession as a physician in
the country. He incurred great additional expense
by keeping up an establishment in London; and was
constantly exposed to much cost from printing,
postage, &c. &c. without the possibility of a return.
His emoluments from vaccine inoculation, contrary
to the glowing anticipations of his parliamentary
eulogists, were not on an average more than £350
per annum; so that there is clear proof that the
gross deficit of capital in the four years immediately subsequent to his removal to London,
amounted nearly to £6000. In compensation for
all this, Dr. Jenner was voted the sum of £10,000
by Parliament, from which were deducted, in the
shape of official fees, &c. nearly £1000, without
taking into account the tedious delay in the payment.

Under such circumstances his friends thought
it necessary that he should again repair to London. But before we follow him thither, it may be
proper to notice the contemporary advancement of

vaccination abroad, and the opposition it encountered at home.

Excepting in the British metropolis, and some of the large provincial towns, no formidable interruption occurred to its progress. Almost all the communications from foreign countries were gratifying in the extreme. Dr. Frank, who had recently gone to Wilna as Professor of Pathology, on the 13th of January announced to Jenner that that university, wishing to confer a distinguished mark of its esteem, had chosen him an honorary member, and transmitted the diploma. About the same time, Dr. Barboza informed him of the successful progress of vaccination in the Brazils; all the civil authorities assisting his efforts. The manner in which he procured his vaccine lymph deserves to be recorded; it shows an energy in the Brazilian government highly creditable. He had carried vaccine matter from England in 1803. He reproduced it in Lisbon; but it failed when carried across the Atlantic. It was therefore resolved to send some boys to Lisbon, who were successively vaccinated on their homeward passage.

A letter from Dr. Scott of Bombay contains intelligence equally cheering regarding the continent of Asia. The stock of vaccine virus had been kept up with perfect success. About the same period Dr. De Carro transmitted intelligence from other parts of our Indian possessions. He had

already received authentic information of upwards
of 35,000 persons being vaccinated in the Mysore.
This letter, amidst a great deal of valuable and interesting matter, contains a statement from Dr.
Auban of Constantinople, which seems to prove
that the Arabians had been long acquainted with
cow-pox. This rests on the authority of Dr. Gem-
mini of Constantinople, who found in an Arabian
MS. written five or six hundred years ago, an account of this affection of the cows. The same
work also notices an eruptive disease of sheep.
I know not where to find any account of this Arabian MS., but the facts, so far as they go, confirm
and illustrate the historical evidence touching the
eruptive diseases of the inferior animals given in
the preceding volume. Dr. De Carro continued to
prosecute his investigations on these subjects with
great zeal and intelligence. He suggested to all
his correspondents in the East inquiries respecting the diseases of horses and cows particularly.
One of his friends, Mr. Barker, the British Consul
at Aleppo, gave him very valuable intelligence
with regard to a disease prevalent amongst the
horses of Arabia, which seems to spread through
the stud, much in the same way that cow-pox does
through the dairies in England.

Tidings were likewise received of the introduction
of vaccination into the Isle of France from the
Coromandel coast. In Bengal, notwithstanding the
encouragement afforded by government, it was

found very difficult to overcome the obstacles
against the practice. This year, however, the chief
of the Mahratta Empire, Dowrat Row Scindia, ordered his only child to be vaccinated. " This sensible
act of the head of the Mahrattas (says Dr-
Shoolbred in his letter to Dr. Jenner, announcing
the fact), who are all Hindoos of high caste, shews
that it is on no essential point of religion that they
object to the new practice. The Brahmins, it appeared, who are in the habit of inoculating for
small-pox, and were interested in continuing the
practice, contrived to excite prejudices in the minds
of the people, on grounds which it was imagined
at the outset would have facilitated the adoption
of it. Its origin from the cow, instead of impress-
ing them in its favour, as was supposed it would
do, was converted by the Brahmins into an argument against its use, as they contended it was
thereby rendered impure to them."

The alleged power of vaccination in controlling
the plague, attracted at this period a considerable
degree of public attention. Jenner received letters on the subject from many quarters. The circumstance having been made known to the
Spanish Consul at Morocco, a communication was
transmitted to Sir Joseph Banks on that subject,
and from him to Jenner. His sentiments on this
point were thus communicated to another correspondent.

" I never was so sanguine in my hopes of seeing

the plague extinguished by vaccine inoculation as some of my friends were, as you may have seen by the few introductory lines which accompanied Dr. De Carro's letter in the public papers." " I will just drop a hint : the vaccine disease, in my opinion, is not a preventive of the small pox, but the small-pox itself; that is to say, the horrible form under which the disease appears in its contagious state is (as I conceive) a malignant variety. Now, if it should ever be discovered that the plague is a variety of some milder disease generated originally in a way that may ever elude our research, and the source should be discovered from whence it sprang, this may be applied to a great and grand purpose. The phenomena of the cow-pox open many paths for speculation, every one of which I hope may be explored."

From about 1804, the reports of failures in vaccination had begun to multiply. The fears of some of his friends had been thereby excited to rather an immoderate degree. Jenner certainly deplored the ignorance that gave occasion to such rumours, but he felt no anxiety concerning his great and fundamental position. Writing to the late Lord Berkeley on this subject he says, " I expect that cases of this sort will flow in upon me in no inconsiderable numbers ; and for this plain reason—a great number, perhaps the majority, of those who inoculate are not sufficiently acquainted with the nature of the disease to enable them to

discriminate with due accuracy between the perfect and imperfect pustule. This is a lesson not
very difficult to learn, but unless it is learnt, to
inoculate the cow-pox is folly and presumption."

Another correspondent, who was seriously
alarmed for Jenner's reputation, wrote him a long
letter full of doleful anticipations of the ill effects
likely to arise from the sinister rumours propagated by the anti-vaccinists, and advised him to
come forward and vindicate his doctrines. This
was the response.—" The post is just come in,
and I have been entertaining Mrs. Jenner and
my family with your dream. Some kind friend
had perhaps thrown your stomach into disorder
by tempting you to go too deep into an oyster-
barrel; or had our friend P—— seduced you
with the fumes of one of his favourite supper
dishes? A devil, or a something, had certainly
disordered your stomach ; and your stomach
shewed its resentment on your head; and your
letter is the consequence. However, I will reason
on it for a moment as if it were not a dream.
You are imposed upon, and so is my friend Fox.
Vaccination never stood on more lofty ground
than at present. I know very well the opinion
of the wise and great upon it; and the foolish
and the little I don't care a straw for. Why
should we fix our eyes on this spot only ? Let
them range the world over, and they must contemplate with delight and exultation what they

behold on the great Continents of Europe and America ; in our settlements in India, where all ranks of people, from the poor Hindoo to the Governor-General, hail Vaccina as a new divinity. In the island of Ceylon my account states that upwards of thirty thousand had been vaccinated a twelvemonth ago. I could march you round the globe, and wherever you rested you should see scenes like these. *There* I have honour, *here* I have none : and let me tell you, whatever my feelings may have been on this subject, they are now at rest. What I have *said on this vaccine subject is true. If properly conducted, it secures the constitution as much as variolous inoculation possibly can. It is the small-pox in a purer form than that which has been current among us for twelve centuries past."*

" You and my city friend suppose me idle—that I no longer employ my time and my thoughts on the vaccine subject. So very opposite is the real state of the case, that were you here (where I should be very glad to see you), you would see that my whole time is nearly engrossed by it. On an average I am at least six hours daily with my pen in my hand, bending over writing-paper, till I am grown as crooked as a cow's horn and tawny as whey-butter ; and *you* want to make me as mad as a bull: but it won't do, Mr. D. ; so good night to you. I'll to my pillow, not of thorns, believe me, nor of hops; but of poppies, or at least of something that produces calm repose."

Dr. Moseley's conjectural arguments against vaccination have been already mentioned. The next person of any name who appeared on the same side of the question was Mr. Goldson of Portsea. He was a man of some character in his neighbourhood; and this circumstance, together with the title of his pamphlet and the important disclosures which it was said to contain, gave it much more consequence than it really merited. He was evidently a prejudiced witness, and almost as ignorant of the subject on which he wrote as his great prototype. It is to be feared, likewise, that he allowed personal feelings to interfere with his judgment. He complained that Dr. Jenner had slighted a private communication of his; but even if this had been true (which it certainly was not), it would not justify hasty and inaccurate statements, or hostile representations concerning a practice which had for its sole object the mitigation of human misery. Jenner never permitted any irritation which might have arisen from occurrences of this kind to interfere with his own conduct. At a subsequent period he wrote to Mr. Goldson, and endeavoured to remove his errors. He even invited him to his house, hoping by personal intercourse and discussion to explain any points which he evidently did not understand. All these overtures were rejected; but whether Mr. Goldson continued in his opposition to vaccination, I cannot tell.

I thought it worth while to mention these incidents, not, certainly, from the importance of Mr.
Goldson's pamphlet itself, but because the discussion which arose from it was really of considerable
interest, and brought out many facts which ought
to have been remembered when the reputed
failures of vaccination attracted so much notice
some years ago. He was quite wrong himself,
and his vaccinations were evidently imperfect;
but the possibility of failure, and the circumstances
under which it may occur, drove then to examine
more closely the history of small-pox, as well as
the affinity between that disease and cow-pox;
and had the information which was then acquired
been duly considered, some recent occurrences of
an unsatisfactory nature might have been easily
solved.

In former chapters I have endeavoured both to
shew how Dr. Jenner felt on this subject, and to
corroborate his opinions. The only successful
attempts that have been made to explain the
nature of the Variolæ Vaccinæ have proceeded
from those who have followed his steps. Some
injudicious friends of vaccination were at great
pains to impress the public mind with other sentiments; and represented vaccination as so different in all respects from small-pox inoculation as
to render any attempt to trace affinities between
them unphilosophical and absurd. The cause they
meant to serve by this conduct was not advanced

by it. Had they been accurately acquainted with
the past history of small-pox, such reasoning would
never have been adopted. It has been fully shewn,
that in the inoculation for that disease, the mitigation of the symptoms depended very much on
the number of pustules produced; and that even
one pustule could afford protection from subsequent attacks. To moderate the eruption and to
subdue the fever was the object of all the enlightened physicians who treated small-pox. Inoculation most materially aided these intentions;
vaccination carried them into full effect. So complete, indeed, and so wonderful, were the results of
this practice, that all past experience was neglected
or overlooked; and notions of a very extravagant
kind were adopted, and the facts themselves,
which had been accumulated, were rendered less
convincing, if not altogether inconclusive, by being
mixed up with false theory.

It ought ever to be borne in mind, that the
laws which govern the Variolæ Vaccinæ were discovered by Jenner through the medium of his
knowledge respecting small-pox. The same knowledge enabled him to trace the deviations in the
former; and the same kind of information has
supplied us with the only means of explaining the
difficulties that recent epidemics have presented.

The year 1804, in his estimation, formed an era
in the history of the Variolæ Vaccinæ. The
assertion, that the cow-pox afforded only a tem-

porary security was then insisted on. Had it
been correct, it would have deprived the discovery
of nearly all its value. This assertion was very
easily made; and, in the infancy of the practice,
could not be well disproved. To these circumstances it was owing, that the crude and unsupported statements of Mr. Goldson acquired any
influence. Dr. Jenner himself, from the commencement, perceived that in his cases of failure,
cow-pox had never properly taken place.

The real merits of the question were also detected by Mr. Dunning. This venerable individual was, as I have already mentioned, one of the
first British surgeons who stood forward to recommend vaccination soon after the practice was promulgated; and from that period to the present
time he has upheld the accuracy and justness of
Jenner's views. His little tract, published about
this time, under the title of, " A short Detail of
some circumstances connected with Vaccine Inoculation," &c. &c. contains some of the soundest
opinions with regard to the nature of Variolæ and
of Variolæ Vaccinæ that have ever appeared. The
steps by which he was enabled to extricate himself from the difficulties of new and perplexing
occurrences, were the same that conducted Jenner through the mazes of his first investigation,
and continued to guide him to the last.

It was by studying small-pox that he became
thoroughly acquainted both with the benefits con-

ferred by vaccination, and the principles that
ought to direct the practice. His mind seems to
have harmonized peculiarly with Jenner's on all
these points. He was not a servile imitator ; the
same spirit of knowledge and of truth which enlightened Jenner, seems to have been at once
communicated to Mr. Dunning. There was a
moral sympathy, as well as an intellectual affinity,
which enabled him, on the instant, to perceive and
to enter into the peculiarities of the character and
the value of the discoveries of Jenner.

These statements are supported by the experience of nearly five and thirty years, during which
time the efficacy of cow-pox has been put to the
severest trials. The result of the whole has shewn,
that all who cordially and honestly embraced the
opinions of Jenner, have advanced without waver-
ing or uncertainty ; and have found by ample testimony, that in so doing they had chosen a safe
leader. Some of the points which have been matter of discussion even within these few years, were
very fully explained at the period just referred to.
It was then clearly ascertained, that there were
deviations from the usual course of small-pox,
which were quite as common, and infinitely more
disastrous, than those which took place in vaccination. These deviations regarded two apparently
different states of the constitution. In the one, the
susceptibility of small-pox was not taken away by
previous infection ; while, on the other hand, some

constitutions seem to be unsusceptible of small-
pox infection altogether. It was found, that similar occurrences took place in the practice of vaccination; but as the security which the latter
afforded was more likely to be interfered with by
slight causes than the former, it became absolutely necessary that great care should be shown
in watching the progress and character of the
pustule. Dr. Jenner had from the beginning felt
the propriety of this watchfulness; and had dis-
tinctly announced, that it was possible to propagate
an affection by inoculation conveying different degrees of security, according as that affection approached to, or receded from, the full and perfect
standard. He also clearly stated, that the course of
the vaccine pustule might be so modified as to deprive it of its efficacy; that inoculation from such
a source might communicate an inefficient protection, and that all who were thus vaccinated were
more or less liable to subsequent small-pox. His
directions for obviating occurrences of this kind
regarded, *first*, the character of the pustule itself,
the time and quality of the lymph taken for inoculation, and all other circumstances that might
go to affect the complete progress of the disorder.
He attached great importance to this last point;
and in the course of this year published his tract,
" On the Varieties and Modifications of the Vaccine Pustule, occasioned by an herpetic state of
the Skin." I cannot refer to this publication with-

out calling the attention of my reader to the following sentence in the introduction : " I shall here just observe, that the most ample testimonies now lie before me, supporting my opinion that the herpetic, and some other irritative eruptions, are capable of rendering variolous inoculation imperfect, as well as the vaccine."

Besides the instructions which Dr. Jenner himself had published, for the purpose of securing perfection in the vaccine process, Mr. Dunning has the merit of establishing a canon, which is now, I believe, universally adopted, namely, that one pustule *at least* should remain undisturbed. Dr. Jenner most candidly admitted the propriety of Mr. Dunning's remarks. In a letter to that gentleman, dated July 29th, 1805, he says, " From what you have already said and observed on the subject, I inculcate every where the propriety of observing greater precaution, and I found it entirely on your observations in conducting the process of vaccination. I recommend invariably two pustules, and that one should remain unmolested."

With equal candour he gives his testimony to the accuracy of Mr. Dunning on another point which bears upon this question. In a letter to Dr. Willan, dated Feb. 23, 1806, he says, " It strikes me that the constitution loses its susceptibility of small-pox contagion, and the capability of producing the disease in its ordinary state, in proportion to the degree of perfection which the vaccine vesicle has put on in its progress; and that the small-pox, if taken subsequently, is modified accordingly." And he then adds, " This opinion was first published by Mr. Dunning of Plymouth."

It is of considerable consequence to recall these facts at this time, because they may, perhaps, teach a lesson of practical wisdom which ought not to be lost sight of. The information then acquired, doubtless might have prevented many of the disasters which have since occurred, had those who were chiefly concerned studied vaccination as it merited.

As a commentary on the preceding remarks, I subjoin a long and valuable letter written by Dr. Jenner to Mr. Dunning. For the same reason I mean to put on record several other extracts illustrative of his doctrines. They had not fallen into my hands when the former volume of this work was preparing for the press, otherwise they would have formed most interesting additions to the train of argument which was then put forth with respect to the nature of the Variolæ Vaccinæ.

Berkeley, Dec. 23, 1804.

My dear Friend,

I thank you for your obliging attention. Your communication of this evening affords me great pleasure; I mark in it the determination of wise and good men to overcome that prejudice and obstinacy which has kept afloat the fatal poison in your neighbourhood. My best wishes will ever attend the philanthropic *nine*.

Foreigners hear with the utmost astonishment, that " in some parts of England there are persons who still inoculate for small-pox." It must, indeed, excite their wonder, when they see *that* disease in some of their largest cities, and in wide-extended districts around them, totally exterminated. Let us not, my friend, vex ourselves too much at what we see here; let us consider this but a speck, when compared with the wide surface of our planet, over which, thank God! Vaccina has every where shed her influence. From the potentate to the peasant, in every country but this, she is received with grateful and open arms. What an admirable arrangement was that I sent you in my last letter, made by the Marquis Wellesley, the Governor General of India, for extermination of the small-pox in that quarter of the globe, contrasted with our efforts here! What pigmies we look like! Did you see the Quarterly Report of the Royal Jennerian Society? It was published in some of the evening papers a few days ago. As far as regards the progress of vaccination in the metropolis, and its influence on the mortality occasioned by the small-pox, it is very good; but how shameful to see a society, constituted for such a purpose, and of which the Royal Family of England bears a part, begging a few guineas of the community for the support of its expenses! This is literally the case, while horses are protected from diseases and death by national munificence *. Shame on it! I must drop the subject, or I shall grow as warm as my friend John Ring.

You speak of Ring and Goldson. Recollect there was not time to be cool. What lover of vaccination—what man well acquainted with its nature, and that of the small-

* This refers to the establishment of the Royal Veterinary College.

pox, could read Goldson's book, and lay it down coolly?
Ring, the moment he read it, and that, indeed, which was
infinitely worse than the book itself, the murderous har-
binger—the advertisement, instantly charged his blunder-
buss, and fired it in the face of the author. I must freely
confess, I do not feel so cool about this Mr. Goldson, as
you do. His book has sent many a victim to a premature
grave; and would have sent many more, but for the
humanity and zeal of yourself and others who stepped
forward to counteract its dreadful tendency. Had Goldson
but written a simple letter to me, stating those occurrences
in his practice which appeared extraordinary, I should
with the greatest pleasure have told him where the mistakes lay, and made him a good vaccinist. By the way,
it has been represented to me, that Mr. Goldson once
wrote to me, and that I did not answer his letter. This
is not true. He wrote a very civil letter to me during the
sitting of the committee on the affair of Clarke the marine,
so malignantly taken up by Hope. This letter I answered
almost immediately; and inclosed one of my papers of
instructions for vaccine inoculation. I am almost certain,
too, my letter was franked by the chairman of the committee, Admiral Berkeley; and now respecting that point,
which seems to give you so much uneasiness—Hitchins's
child. In my last letter I really gave you the result,
though it lay in a small compass, of my deliberations.
In such a case, there is nothing but the fact before you.
Conjecture may be endless as to the cause and consequence; with regard to the latter, I ventured to go a little
way by surmising, that the mild innoxious small-pox,
which appeared in this country some years ago, might
have had its origin from a case similar to that of Mr. Hitchins's child. There is a medical gentleman, surgeon of
the South Gloucestershire Militia, who tells me, that he

is so susceptible of the contagion of the small-pox, that he
never attends a patient with that disease without catching
it, evident marks of which appear upon his skin.   Mr.
Shrapnell's countenance shews the signs of his having had
the disease rather violently in the first instance.   As we
have such abundant testimonies of persons having the
small-pox a second time, why may there not be exceptions to the cow-pox giving security in every instance of
inoculation *?   The number of our inoculations is now
incalculable; yet how few have been the exceptions.
When they happened in variolous inoculation, there was
no one ready to come forth and blow the hostile trumpet.

But to return.   There may be peculiarities of constitution favourable to this phenomenon.   My opinion still is,
that the grand interference is from the agency of the
herpes, in some form or another; for I have discovered
that it is a very Proteus, assuming, as it thinks fit, the
character of the greater part of the irritative eruptions
that assail us.   I shall have much to say on this disease
one of these days.   My paper in the Medical Journal
seems not to have excited the smallest interest, though I
venture to predict it will be found highly interesting.
There is no getting a line from your medical neighbour,

---

* The child of a cousin of mine, who was vaccinated in
India, and apparently with success, had the operation repeated after he arrived in England, and again received the
infection.   This child was subsequently inoculated for the
small-pox, and received the disease.   But this is not all;
he was recently exposed to the influence of this contagious
disorder, and took it in a casual way.   Could a stronger illustration be adduced of the doctrine laid down in these volumes
touching the identity of the diseases in question, with relation to their protecting power?—J. B.

Mr. Embling. Will you be so good as to send him the inclosed note?

I cannot conclude my letter, long as it is already, without telling you that I have lately received from the surgeon-general, Mr. Christie, in the Island of Ceylon, a most charming account of the progress of vaccination there. Notwithstanding the impediments arising from a state of warfare, upwards of thirty thousand persons had been vaccinated. The desolation which the small-pox occasioned when it broke out among the inhabitants, almost exceeds credibility. Variolous inoculation proved so unfavourable, that the people would not be prevailed upon to receive it; indeed, that must be the same all the world over. Where it is put in practice, the disease must necessarily spread. Of this truth you have lately seen some examples. I am much pleased to find that the resolution of the worthy *nine* is going to the papers. If these were better guarded, the enemy would not make such frequent irruptions.

Believe me, with great respect,

Yours very faithfully,

EDWARD JENNER.

EXTRACTS OF LETTERS FROM DR. JENNER TO
MR. DUNNING.

July 5th, 1804.

There is not a single case, nor a single argument, that puts the weight of a feather in the scale of the anti-vaccinist. That which seems to be the heaviest, becomes light as air, when we consider that the human constitution is at one time susceptible of variolous contagion, at another, not so; and this insusceptibility sometimes continues to a late period of life. Elizabeth Everet was a small-pox nurse in this neighbourhood for forty years. She supposed she had had the small-pox when a child. A few years since she was sent for to Bristol to nurse a patient, caught the disease, and died.

Mr. Long, surgeon of St. Bartholomew's, had a similar instance in his own family.

A thousand thanks to you and Dr. Remmet for your investigation of the Exmouth case. Never mind; you will hear enough of small-pox after cow-pox. It must be so. Every bungling vaccinist who excites a pustule on the arm will swear like G. it was correct, without knowing that nicety of distinction which every man ought to know, before he presumes to take up the vaccine lancet.

March 1st, 1805.

The security given to the constitution by vaccine inoculation, is exactly equal to that given by the variolous. To expect more from it would be wrong. As failures in the latter are constantly presenting themselves, nearly from its commencement to the present time, we must expect to find them in the former also. In my opinion, in either case, they occur from the same causes; one might name for example, among others, some peculiarity of constitution which prevents the virus from acting properly, even when properly applied; from inattention, or want of due knowledge in the inoculator; particularly in not being able to discriminate between the correct and incorrect pustule.

March 9th, 1805.

Think for a moment of my situation before you censure me for tardiness—the correspondence of the world to attend to. The pressure is often, I do assure you, so great, that it is more than either my body or mind can well endure. You say, "let vaccination, for God's sake, rest on its own foundation." My dear sir, that is exactly what I want, and the course I have been pursuing. Neither the impudence of Pearson, the folly of Goldson, nor the baseness of Moseley and Squirrel, to which I may add the stupid absurdity of Birch, has put me out of my way in the least,—and why? I placed it on a rock, where I knew it would be immoveable, before I invited the public to look at it.

Extracts from a Letter of Dr. Jenner's to Dr. Evans, Ketley-Bank.

How little he (Mr. Cartwright) must have known of the agency of variolous matter, to have argued as he has done. Wonderful as it is, yet there are abundant facts to prove, that the insertion of variolous matter into the skin has produced a virus fit for the purpose of continuing the inoculation; and yet the person who has borne it, and on whose skin it was generated, has subsequently been infected with the small-pox, on exposure to its influence. Just so with the vaccine.

Again, in the same letter, he (Dr. J.) adds :— "Vaccine inoculation has certainly unveiled many of the mysterious facts attendant upon the small-pox and its inoculation. How often have we seen (apparently) the full effect on the arm from the insertion of variolous matter, indisposition, and even eruptions following it, and its termination in an extensive and deep cicatrix; and yet, on exposure, the person who underwent this, has caught the small-pox. Your practice in the populous neighbourhood you reside in, was formerly, I believe, very extensive in variolous inoculation. Did you ever perceive any connexion between the state of the skin and the progress of the pustule on the arm? I can scarcely flatter myself with the hope of finding much accurate information on this point, as the *arm* became a secondary object in inoculating the small-pox, our solicitude being directed to the number of pustules."

In one of his Journals he has left the following notes upon the same subject.

"The origin of small-pox is the same as that of cow-pox; and as the *latter* was probably coeval with the brute creation, the former was only a variety springing from it."

"There are certainly more forms than one, (without considering the common variation between the confluent and distinct) in which the small-pox appears in what is called the natural way."

"It will be inquired (if the foregoing reasoning be *à priori* correct) in what way can the action of cow-pox (or the equine pock) in preventing subsequent small-pox be reconcilable with the established laws of the animal economy? My reply is, for the reasons which I have stated on the basis of facts, that they were not *bonâ fide dissimilar* in their nature; but, on the contrary, *identical*. On this ground I gave my first book the title of 'An Inquiry into the causes and effects of the *Variolæ*

Vaccinæ '—a circumstance which has since been regarded by many as the happy foresight of a connexion, which was destined by future evidence to
become more warranted."

How admirably subsequent investigations have
confirmed these most sagacious remarks !

The Medical Society of London having voted
Dr. Jenner a gold medal in honour of his discovery,
he was invited by the President, Dr. Sims, to attend
their anniversary festival on the 8th of March, on
which day an oration is usually delivered. The
gentleman appointed to this office had unfortunately been taken ill, and could not attend. In
this dilemma the Society applied to Dr. Lettsom,
who chose for his subject the Jennerian discovery,
and the delivery of the gold medal. He pronounced
a very eloquent and impressive discourse, which
was afterwards printed in the European Magazine
for 1804. Dr. Jenner could not appear personally
to receive this mark of respect. Dr. Lettsom, however, very ably supplied his place in the Metropolis. He likewise fought his battles, and often sig-
nally vanquished his opponents. An account of a
scene of this kind is very descriptive, and not
without interest from the character of the parties
concerned.

August 1st.

Although after writing the preceding letter, *super strata
viarum*, I arrived home soon enough to catch the post,

I was so pulled away professionally that I had not time
to seal it. I received an invitation to dine with a party;
but could not attend till past eight. When arrived, I
found Mr. Alexander, M. P. Chairman of Ways and
Means; the Bishop of Cloyne, the Rev. Dr. Parr, Dr·
Pearson, Dr. Shaw of the British Museum, Mr. Planta of
the Royal Society, Rev. Mr. Maurice, author of Indian
Antiquities. Somehow, Pearson introduced the House of
Commons—their Committees—when he made a Philippic
of half an hour's abuse of that Committee which recommended Dr. Jenner's discovery, as a *rascally*, ignorant
business for what was no discovery, and concluded with
severe animadversions on Dr. Jenner. After he had
finished, Dr. Parr seemed persuaded of Dr. Jenner's un-
worthiness; and Mr. Alexander said, had it depended upon
his casting vote, as he was Chairman, he would have given
it against Dr. Jenner. I then requested to be heard for
an absent friend. I went over the whole ground with a pers-
picuity I never possessed before. I exposed the conduct
and mistakes of Pearson and Woodville in so strong a
manner, that after listening to me half an hour, every person seemed electrified but Pearson. One divine started
up, took me by the hand, clapped me on the back, and
embraced me, and declared that I had incontrovertibly
proved Jenner, not merely the promulgator, but inventor
and discoverer. Parr exclaimed, I would have voted Jenner ten times ten thousand pounds. Alexander declared
that he now saw the matter in a new and convincing point
of view. Pearson then made a reply of above half an
hour, and when he had concluded, Dr. Parr was appointed
to decide upon the facts; and these were his words—" Dr.
Lettsom has *convinced* me that Dr. Jenner is the discoverer, and Dr. Pearson's defence has *confirmed* me in that
conviction." 1 asked Dr. Pearson if he had any thing more

to say. He said he had done, and now I trust he will never again venture, at least in the presence of the company, or any one of them, to broach his unfounded invectives ; and I think he has now received his quietus. He little expected that I could have explained his mistakes and Woodville's so clearly. I mentioned facts which thunderstruck him, of which even you are ignorant, respecting this base coalition against you. He seemed confounded. His friend Maurice and his devotee ran about the room—" Lettsom has conquered, Lettsom has conquered." Parr said he would come and see me, and Mr. Alexander proposes me the same honour. I know that Maurice will talk of this rencounter every where.

J. C. L.

Jenner continued to receive from many public bodies marks of distinction, all which he valued most highly, not only because they were grateful to his own heart, but because they materially contributed by the sanction attached to them to extend the practice which he had the happiness to discover. In this spirit he obtained the intelligence of a degree conferred on him by the Harvardian University of Cambridge, in Massachusetts. The Diploma was transmitted by his friend Dr. Waterhouse, and it arrived in England during the spring of 1805.

The Corporation of Dublin, about the same time, unanimously voted him the freedom of that city. In announcing this to Dr. Jenner, the officers of that respectable civic body transmitted a charge

of somewhat about five pounds sterling for his admission fees.   This mode of making him open his
purse strings for a gratuitous honour used often to
excite a good-natured smile on his countenance
when he adverted to the transaction.

In mentioning these honours, it is gratifying to
observe that they proceeded not only from persons
devoted to science, but from men associated together of every denomination, from the municipal
authorities of imperial cities to the humbler cor-
porators of smaller towns.   The magistrates of
Edinburgh took the lead, in this respect, of its own
College of Physicians, and voted him the freedom
of their ancient metropolis.   The City, in announcing this honour to Dr. Jenner, availed itself of the
kindness of one of the most venerable and distinguished of its learned Professors, Dr. Andrew Duncan, senior.   I cannot mention this individual without expressing my thankfulness for his goodness in
directing my early studies.   His indefatigable industry, his active benevolence, and his great acquirements, must cause his name to be respected by
every enlightened physician.   He was one of the first
to promote the practice of vaccination in Scotland.
In doing so, he illustrated that principle which
guided him in all his duties.   He was constantly
alive to the progress of scientific truth, and omitted
no opportunity of diffusing it.   Since these remarks were written, I have received the melancholy tidings of the death of this estimable and

venerable man. All my early professional recollections are associated with him as my adviser and
instructor; and I should be guilty of deep ingratitude were I not to seize this opportunity of announcing my past obligations for his kindness and
generosity, and my unfeigned respect for his memory.

In 1804, one of the most beautiful of the Napoleon series of medals was struck commemorative of
the Emperor's estimate of the value of vaccination.
It has been said, that it was at the same time intended as a mark of personal honour to Jenner, by
appropriating one side of the medal to his bust.
Whether this really was the design, I cannot say;
but it is certain that the obverse of some of the
early medals was left blank. Subsequently it was
occupied by the head of the Emperor.

He who flushed with victory and at the head of
the revolutionary army of France had spared the
university of Pavia, out of respect to the genius
of Spallanzani, when the city itself was given up
to plunder, proved that the claims of science were
not forgotten amid the astonishing events which
carried him forward to the highest pinnacle of am-
bition. His animosity to England had been shown
in that vehement and decided manner which
marked all his actions; yet there was one chord
of sympathy unbroken, and which, when duly
touched, showed that his intoxicating success had
not raised his proud spirit beyond some of the

calls of justice and humanity, and that he could still be moved by the peaceful arguments of truth and science.

His unjust detention of the inhabitants of Great Britain, who were quietly sojourning in his dominions after the peace of Amiens, was in direct opposition to the usages of modern warfare. Dr. Jenner's efforts to release some distinguished individuals who had been so detained, have been already mentioned. He was now called upon to exert himself in a different way, and his efforts brought him more immediately into contact with Napoleon himself. Dr. Wickham, who was one of the travelling fellows of the University of Oxford, was at Paris in the first years of his travels, when the command for the detention of the English was issued. He was permitted to retire to Geneva on his parole. Another young Englishman, of the name of Williams, was also detained. The situation of both of these gentlemen had been particularly submitted to the consideration of Dr. Jenner: that of Dr. Wickham presented peculiar claims. He was travelling in compliance with the generous purposes of the founder of his fellowship; he was pursuing those objects which in all wars have been held sacred, and on that account alone might have claimed exemption from the ruthless decree. Mr. Williams, too, was travelling for improvement, though not under the immediate sanction and the special commission of a learned body. His

health had suffered materially, and it was an object
of the greatest consequence both to him and to
Dr. Wickham, to be liberated from their harassing
bondage.

Dr. Jenner, in the first instance, directed his
applications for the accomplishment of this object
through the Central Committee of Vaccination at
Paris; but they all failed. Under these circumstances, the Committee recommended that he
should immediately avail himself of his great celebrity in France, and directly appeal to the Emperor himself. He profited by this advice, and
addressed the following letter to his Imperial
Majesty, of which both the French and English
copies have been found among his papers.

DR. JENNER TO NAPOLEON.

SIRE,

Having by the blessing of Providence made a discovery
of which all nations acknowledge the beneficial effects, I
presume upon that plea alone, with great deference, to request a favour from your Imperial Majesty, who early
appreciated the importance of vaccination and encouraged
its propagation, and who is universally admitted to be the
patron of the arts.

My humble request is that your Imperial Majesty will
graciously permit two of my friends, both men of science
and literature, to return to England : one, Mr. William
Thomas Williams, residing at Nancy; the other, Dr. Wickham, at present at Geneva. Should your Imperial Majesty

be pleased to listen to the prayer of my petition, you will impress my mind with sentiments of gratitude never to be effaced.

I have the honour to be, with the most profound deference and respect,

Your Imperial Majesty's

Most obedient and humble servant,

E. J.

Berkeley, Gloucestershire,

February, 1805.

When this letter was despatched, Napoleon was in Italy, but Mr. Williams had an opportunity of delivering a copy of it into his Majesty's hands as he passed through Nancy.*   A duplicate was presented by Baron Corvisart, the Emperor's physician, in the month of June 1806.   Early in the following July Mr. Williams received an intimation from Corvisart that the Emperor had listened to Dr. Jenner's petition, and had granted liberty both to himself and to Dr. Wickham.

Dr. Trotter, who took so active a part in promoting subscriptions for the medal presented to Dr. Jenner by the medical officers of the Navy, dedi-

---

* It was either on this or some similar occasion, when Napoleon was about to reject the preferred petition, that Josephine uttered the name of Jenner. The Emperor paused for an instant, and exclaimed, " Jenner ! ah, we can refuse nothing to that man."

cated to him the " Essay on Drunkenness." After
expressing his thanks, Dr. Jenner observes, " I sincerely hope the mirror you have held up may reflect the face of many a drunkard in such a hideous
shape as may terrify and reform. I think it will
prove of great use to young men launching into
the vice of ebriety. The habitual drunkard—the
man engulfed in alcohol, I fear, will scarcely
be able to get out. I have in the course of my
life known *one* instance, and I think that is all."

Jenner had a peculiarly graceful manner of indicating the respect which he entertained for individuals who had, in any way, exerted themselves
in promoting vaccination. I have already mentioned Robert Bloomfield as having struck his lyre
in favour of that cause. The Poem was dedicated
to Jenner and his brethren of the Royal Jennerian
Society. In writing to the author he says, " I
trust it will be as well received, and gain as high
commendation, as the Farmer's Boy. It need not
obtain more. You must allow me to fix upon some
mark of my esteem. Do me the favour, then, to
accept a silver ink-stand, into which the enclosed
may be converted if you will call upon Rundell
and Bridge, Ludgate Hill, and use my name. I
should like the following plain engraving on it.—
' Edward Jenner, M.D. to Robert Bloomfield.' "

The name of the Reverend James Plumptre,
M.A. is now to be added to the list of those clergymen of the Established Church who advocated the

cause of vaccination from the pulpit. In the year 1805 this gentleman had to preach before the University of Cambridge at St. Mary's. On this occasion he deemed it a fit opportunity to lay before the members of that celebrated seat of learning a scriptural view of pestilence, but particularly of small-pox, together with reflections on the nature of cow-pox. On the third of March of the same year he preached in the parish church of Hinxton, Cambridgeshire, another discourse on the same subject. It was designed for a country congregation; and afforded an opportunity to the preacher of addressing to his parishioners many of those arguments in favour of vaccine inoculation which were well calculated to remove some of the prejudices that had been artfully instilled into the minds of the lower classes.

He took his text for both discourses from the 60th Chapter of Numbers, 48th verse, "And he stood between the dead and the living; and the plague was stayed." As became his high and holy calling, he pointed out the invariable connexion between sin and suffering, and shewed that many of the most signal and afflictive dispensations of Divine wrath came in the shape of disease and death.

This truth, illustrated by the whole history of man, naturally induced him to apply it to elucidate the origin of the most universal pestilence known to our species. The opinions which he expressed

on that subject arose from the belief that the
small-pox was comparatively a new disease, and
that the sudden appearance of it at the time of the
rise of the Mahometan imposture pointed it out as
a visitation foretold by prophecy, and illustrated
the pouring out of the first phial upon the earth
when " there fell a noisome and grievous sore upon
the men who had the mark of the beast, and upon
them which worshipped his image." It would not
become me to pronounce upon the accuracy of this
interpretation. Although the view taken in the
former volume sustains the reverend author's
general position, it, nevertheless, does not accord
with his specific interpretation, inasmuch as it
renders it probable that small-pox first visited our
race at a period long antecedent to that assigned
to it in this discourse. It was elucidated by
copious and useful notes, many of them containing
valuable information of a practical nature, and the
whole written in a manner calculated to instruct
both the divine and the physician. The dedication to Dr. Jenner was singularly appropriate, and
expressed in warm and becoming terms the writer's
veneration and attachment.*

---

* TO EDWARD JENNER, M.D. F.R.S. &c. &c.

" The instrument in the hands of a gracious Providence
of discovering vaccine inoculation, and the disinterested
divulger of that salutary blessing for the benefit of the whole
human race. This discourse, intended to connect the practice of medicine with religion, and to set forth the just wrath

This benevolent clergyman did not confine himself to his efforts in the pulpit. He took active
measures to meet the progress of small pox, whenever it appeared in his vicinity. Sometimes he
employed, at his own cost, a medical man to vaccinate the poor; at others he took that office upon
himself. He likewise printed and circulated largely
songs and ballads calculated to impress the minds
of the lower orders with the benefit of vaccination.*

and power, and more particularly the infinite goodness, of our
Almighty Father, is inscribed as a small token of that veneration and gratitude due from an admiring world, by his obedient and humble servant,

"JAMES PLUMPTRE."
"Clare Hall, Cambridge, Feb. 25th, 1805.

* Since the publication of the former volume, Mr. Plumptre has had the kindness to transmit to me the following
memorandum. It is a curious and not unimportant testimony in favour of the doctrine which I have endeavoured to
establish respecting the nature of the variola and the variolæ
vaccinæ.

"Among my memoranda on vaccination, I find the following: ' In the public library at Lausanne there is a curious
manuscript by St. Maire, the fourth christian bishop of Lausanne, who died A. D. 601, which he calls a chronicle of his
own times. Among other things which this chronicle contains is the account of a visitation of the small-pox, which he
says made great ravages, and he notices particularly that it
proved very fatal to the cows.'

"This memorandum is in the handwriting of my late sister
Mrs. Anne Plumptre, who translated Bertram's account of

Others acted a very different part. A calumny which was directed both against Dr. Jenner's moral and professional character issued from one of our universities ; it was whispered in the courts of our palaces ; circulated by the periodical press ; and I have reason to know that the effect of the misrepresentation is still felt. This, among numberless instances, proves what hard measure truth receives in this world. In order that I may refute this unworthy attempt to injure both vaccination and its author, I will meet it as it comes from the pen of a reverend divine. Dr. Ramsden, rector of Grundisburgh, Suffolk, in a note to a sermon preached before the university of Cambridge on May 15th 1803, states that he heard with his own ears in the physic school of the university, the king's reader of physic say that Dr. Jenner had, after the discovery of vaccination, inoculated his own son with the small-pox. This sermon, together with the note, was reprinted in 1827.

The king's reader on physic, Sir Isaac Pennington, was a violent opposer of vaccination ; and he put forward his statement with a view to prove that Dr. Jenner, though he recommended the

the Plague at Marseilles, and published many other works : but she does not mention where she got it. I do not recollect reading it in your work, nor do I find it by your *Index*, nor on referring back to the work."

I regret to add that this excellent clergyman has not lived to witness the insertion of this in the present volume.

practice to others, was distrustful of it, and had abandoned it in his own family. It was for the purpose of representing Jenner in this light that the statement was uttered, and printed, and circulated throughout the kingdom. I do not blame the rector of Grundisburgh for repeating in 1803 what he may have heard from the king's reader in physic; but before giving a second edition of the tale in 1827, it would not have been unworthy of his station and his calling, to have ascertained whether the information he had heard was founded in truth, or rested on a basis of a different description; whether Dr. Jenner or his friends had published any explanation of the alleged delinquency; and whether, in short, the real state of the case was not in direct opposition to the inferences which were drawn from it.

The facts were these: On the 14th of May, 1796, Jenner vaccinated his first patient, Phipps. On the 12th of April, 1798, he vaccinated his son Robert, together with several other children. It is particularly specified in his first publication that his son Robert " *did not receive the infection.*" He was, therefore, as much liable to the influence of small-pox contagion as if he had never been vaccinated. Under these circumstances, while the infant was with his parents at Cheltenham, the late Mr. Cother of that place came into Jenner's house, and took the child in his arms, saying that he had just left a family labouring under small-pox.

Jenner immediately exclaimed, " Sir, you know not what you are doing. That child is not protected. He was vaccinated; but the infection failed." Believing that the natural small-pox would certainly follow this exposure, he was greatly distressed and alarmed. He had no vaccine matter. He resolved, therefore, to adopt the next best expedient, and immediately had the child inoculated with small-pox virus, preferring the mitigation which that practice affords to the violence and danger which generally accompany the casual disease.

This simple occurrence, when related as it actually took place, so far from leading to the conclusions that were built upon it, did not afford the slightest ground for them. It was a clear case of professional duty ; and, under like circumstances, every medical man would have been called on to act as Jenner did. He had no vaccine matter; his child was exposed to small-pox contagion ; and what, therefore, did he do ? Small-pox, in some shape, seemed inevitable ; and he sought for that abatement of its virulence which inoculation is known to afford.

At the time the deed was perpetrated, every one who knew the truth was so perfectly satisfied with the soundness and propriety of Jenner's decision, that it never was imagined it could be questioned. The bare fact, however, that Dr. Jenner had employed small-pox inoculation in his own

family, after the publication of his work on the
Variolæ Vaccinæ, was an incident too important
to be lost sight of by those who were unfriendly
to him and to vaccination. " Dr. Jenner may say
what he likes about vaccination, but we know for
certain that he has inoculated his own son with
small-pox." Another repeated this statement with
this addition, that he had done so because he
mistrusted vaccination. A third added another
tint to deepen the colouring, affirming that he
knew that Dr. Jenner had abandoned his confidence in vaccination, and the proof is incontestible, as he has inoculated his own child with
small-pox. These stories were passed from mouth
to mouth, and made a considerable impression on
the public mind before he heard of them, and
were unwittingly uttered by a noble lord in Jenner's presence at St. James's, whom, on the instant, he rebuked in the manner that I shall elsewhere state. They afterwards appeared in print,
and were circulated with every malignant interpretation. They were promptly met by an authoritative statement of the facts given above. That
statement appeared in several of the periodical publications ; and I myself was instrumental in sending
it to more than one in the year 1811, when *for
the first time* I heard of this calumnious misrepresentation. Dr. Jenner likewise, both in conversation and in writing, gave the truth as it actually
occurred ; and there was not a man in the king-

dom who had access to the common sources of information, who might not have acquainted himself with the facts.

Is there not, therefore, some reason to complain of a gentleman of respectable station and character, who reprints a most injurious report without having given himself the trouble to inquire whether the king's reader in physic might not have been in error when he proclaimed it in the University of Cambridge more than thirty years ago? I subjoin the letter addressed to myself in reference to this matter.

MY DEAR DOCTOR,

While my embarrassments thicken upon me, 'tis a very pleasant thing to have so able a friend as you to converse with in this way. I am thrown into a little fresh perplexity, by a letter addressed to me in one of the London papers of yesterday, from Mr. Brown of Musselburgh. Pray look it over, and tell me what course I should pursue. Some notice must be taken of it; but if Mr. Brown thinks he shall be able to draw me into a controversy by such a measure as this, in a public newspaper, he will be mistaken. His letter under the veil of candour and liberality is full of fraud and artifice; for he knows that every insinuation and argument he has advanced, have been refuted, both by the first medical characters in Edinburgh and Dublin; and, indeed, by many others. But the mild, gentle, complaisant antagonist, is a character more difficult to deal with than one who boldly shews his ferocity. I shall avoid further comments till you have seen his production.

From my knowledge of some of the Gloucester vaccinators, I am confident the practice has been very heedlessly conducted there. The patients themselves, and
the parents of some of the vaccinated children, have in
some instances been very culpable. I speak this rather
feelingly; as I have vaccinated children myself brought
from Gloucester to this place, whose arms I never inspected
after the operation.

You request me to give you the history of my son's
inoculation. I do it with pleasure, and beg you to make
any use you please of it.

My two eldest children were inoculated for the small-
pox before I began to inoculate for the cow-pox. My
youngest child was born about the time my experiments
commenced, and was among the earliest I ever vaccinated.
By referring to the first work I published on the subject
in the spring of the year 1798, page 40, you will find his
name, Robert F. Jenner, and you will observe it noticed,
that on his arm the vaccine lymph did not prove infec-
tious. It advanced two or three days, and then died
away. In a short time after I was necessitated to go with
my family to Cheltenham for a few months, where I did
not think it prudent to resume my operations, from a
supposition that the people assembled at a public watering
place might conceive the disease (then so little known)
to be contagious, and that it might excite a clamour.
However, during my stay there, this boy was accidentally
exposed to the small-pox, and in such a way as to leave
no doubt on my mind of his being infected. Having at
this time no vaccine matter in my possession, there was
no alternative but his immediate inoculation, which was
done by Mr. Cother, a surgeon of this place, who is since
dead; but this history is well known to many who are living.

You now see on what a baseless foundation the insinuations which have been published respecting these facts rest.

Believe me,

My dear Doctor,

with much affection,

truly yours,

EDWARD JENNER.

Cheltenham,
6th November, 1810.

CHAPTER II.

CONTINUED PROGRESS OF VACCINATION ABROAD—VACCINA-
TIONS AT CHELTENHAM—ARRANGEMENTSF OR FARTHER
DISCUSSION IN PARLIAMENT—SECOND GRANT—INTER-
VIEW WITH MR. PERCEVAL WITH A VIEW TO LIMIT THE
DIFFUSION OF SMALL-POX BY INOCULATION.

THE hostility in England continued to be happily counterbalanced by multiplied evidence of
the benefits of vaccination from other parts
of the world. A letter from Dr. Friese of Breslau, dated June 9th, 1805, and addressed to
Mr. Ring, commences thus,—" The unremitting
zeal with which you have endeavoured to promote
the Jennerian discovery in your country, and the
interest you have so philanthropically shewn on
hearing of its first providential introduction into
Silesia, will, I hope, excuse me when I take the
liberty to trouble you with some farther account
of the successful progress which that invaluable
prophylactic has since made in this part of the
Prussian dominions." After adverting to the

pamphlets of Messrs. Goldson and Squirrel, the learned writer adds, " At any rate, I am convinced the new doctrine which they promulgate, will find but few proselytes in Germany, where both the governments and the people are more and more sensible of the advantages of the new practice; and where similar equivocal arguments, advanced some years ago by the late Dr. Herz, Mr. Erhman of Frankfort, and Dr. Matterskher of Prague, have been silenced by time and experience.

" You remember, perhaps, by my former letter, that there was also an adversary of some celebrity in Silesia, who rose up against the vaccine inoculation at its first introduction into this country. His name is Mayalla, a physician known in Germany by his very valuable writings on the several mineral waters and bathing-places of Silesia, and by some other works on the veterinary art: but I have the pleasure to inform you, that this respectable practitioner has been converted by *reason* and *evidence* into one of the warmest friends and supporters of vaccination. I must add, that it was particularly by his assistance we are now in possession of two public vaccine institutions at Breslau and Glogau, which are to be regarded as the centres from which the practice is spread, and continues to be spread, through every quarter of the province."

Dr. Friese mentions that, during the past year (1804) the number of vaccinations in all the districts of the Breslau department, had much exceeded those of all the preceding years since 1800. In the department of Glogau more than 10,000 persons were vaccinated. The gross number successfully inoculated with cow-pox, amounted to 34,000 in the two departments above mentioned. Small premiums were given to such poor parents as brought back their children at the proper time for re-examination. Clergymen, likewise, were instructed in the new practice; and exercised it judiciously and skilfully, to the great benefit of their parishioners.

Intelligence equally pleasing was conveyed from other parts of Germany. Jenner received a letter from Mr. B. Levi of London, a gentleman who had just returned from extensive travels in Poland, stating, that in the Russian and Prussian divisions of that country vaccination was making rapid progress; in the former, under the auspices of the Emperor of Russia, and the zealous patronage of the benevolent Empress Dowager Maria, seconded by the indefatigable exertions of Lobenwein, Professor of Anatomy at Wilna. In Austrian Poland, however, the vaccine inoculation was reported to be in a backward state, owing to a very *malignant* kind of *false* cow-pox, propagated by ignorant village matrons and barbers.

By the same conveyance, Dr. Jenner had the satisfaction of receiving accounts from Dr. Frank, Professor at Wilna, whose letter ran thus : " Vac-

cination is thriving apace here; and its progress is
in a very great degree owing to our Professor of
Anatomy, Mr. Lobenwein, whose indefatigable exertions and philanthropy have surmounted many
difficulties naturally to be expected in a country
like this."

Another short extract from this letter of Dr.
Frank will not be unacceptable. " Though so far
distant from London, we did not forget yesterday
(17th of May), to join with the many thousands
of other countries, in celebrating the *anniversary*
of a day so valuable to every well-wisher of science
and philanthropy."

Besides being, as Jenner often expressed it,
vaccine clerk to the world, and attending to
his duties as a physician, his vaccinations were
often most numerous, especially after any alarm
from small-pox in his neighbourhood. He constantly inoculated all who chose to come; and
sometimes he had nearly three hundred persons at his door. Nothing gratified him more
than offices of this kind; and when, in the
course of his practice, any striking proofs of the
efficacy of vaccination presented themselves, his
satisfaction was increased. An occurrence of this
kind happened at Cheltenham, which he used to
relate with great glee. " A poor widow and her
four children chanced to be under the same roof
with a labouring man who had caught the small-
pox. They had been exposed five days to the infection, when an humble neighbour happened to

step in. The poor woman, it appears, had made up her mind to her fate, not seeing the possibility of escape from the calamity that threatened her. However, her wise friend prevailed on her to come to Cheltenham to know what was to be done in such a case; she instantly complied. I happened to be from home; but my servant Richard, who has lived with me many years, exercised his judgment very properly. He soon found out an arm with a fine eighth day pustule, and inoculated the whole group. They have since all been with me full of rejoicing at the consequence. All escaped the contagion except one of the children, on whom appeared a few scattered pocks, or rather pimples, for they did not exceed hemp-seeds in size; nor was the eruption attended with any perceptible indisposition. I have frequently before this disarmed the small-pox of its power on those who had been exposed three days to its contagion; but this fact, with all its circumstances, I own delighted me."

We have another picture of his feelings at the same period in a letter to a friend. " I sincerely hope that you are well and happy, and so does my dear Mrs. Jenner, who, thank God, though not in high health, has gone on better than I once could have expected. As for myself, I bear the fatigues and worries of a public character better by far than those who know the acuteness of my feelings could have anticipated. Happy should I be to

give up my laurels for the repose of retirement,
did I not feel it to be my duty to be in the world.
I certainly derive the most soothing consolation
from my labours, the benefits of which are felt the
world over ; but less appreciated, perhaps, in this
island than in any other part of the civilized
world."—" Cheltenham is much improved since
you saw it. It is too gay for me. I still like my
rustic haunt, old Berkeley, best ; where we are all
going in about a fortnight. Edward is growing
tall, and has long looked over my head. Catha-
rine, now eleven years old, is a promising girl ;
and Robert, eight years old, is just a chip of the
old block."

In order to prosecute the object referred to in
the former chapter, relative both to his present
affairs and the progress of vaccination, he left
Berkeley on the 9th of May 1805, and arrived in
London on the 10th. On the following day he had
an interview with Lord Egremont. On this occasion different measures were canvassed. They all
referred to the establishment of vaccination and
the advancement of his private fortune, which had
been so much injured by what promised to be
beneficial to him. The result of the whole was,
that another application should be made to Parliament. This application was brought about in
the following manner : " During my residence in
town," he observes, " in the summer of 1805, Lady
Crewe happened in conversation to tell me how

much Lord Henry Petty wished for a conference
with me on the vaccine subject; and that he
would like to bring us together. We met at her
villa at Hampstead; and went so fully into the
matter, that his Lordship, convinced of the injury
I had sustained, expressed his determination to
bring something forward in the ensuing session.
Before this session arrived, Mr. Pitt died; and
Lord Henry Petty became Chancellor of the Exchequer. In the early part of the present year
(1806), I again saw his Lordship; and found that
his ardour in my cause had suffered no abatement.
This was soon after proved by his Lordship's motion
in the House."

That other great personages took an interest in
this matter may be gathered from what follows:
" I had not forgot your kind interest about Jenner.
I spoke to the Duke, the Prince, and Morpeth,
and they will all do what you think best; but Morpeth has undertaken to make inquiries whether it
is not possible to bring it again before Parliament.
He thinks if that could be done, it would be more
satisfactory than any subscription. I desired him
to find out how Mr. Pitt was *really* inclined on the
subject, and I only waited the result of these inquiries to write to you *."

On the second of July 1806, Lord Henry Petty
brought the subject of vaccination before the

* Extract of a letter from the Duchess of Devonshire to
J. J. Angerstein, Esq.

House of Commons. This measure was particularly called for at this period, both on Dr. Jenner's account, and the cause of vaccination itself. Its progress had been much obstructed in our own country in consequence of the numerous prejudices which had been excited against it; and small-pox was again becoming prevalent. Such being the case, his Lordship thought that the affair demanded the most serious attention of the legislature. He therefore proposed, that an address to his Majesty should be voted by the House, praying " that his Royal College of Physicians be requested to inquire into the progress of vaccine inoculation, and to assign the causes of its success having been retarded throughout the United Kingdom, in order that their report may be made to this House of Parliament; and that we may take the most proper means of publishing it to the inhabitants at large."

" If," continued his Lordship, " the result of such proposed inquiry turn out (as I am strongly disposed to think it will), a corroboration of the beneficial effects which other nations seem convinced are derived from vaccine inoculation, it will satisfy the people of this country of the many evils which arise from the rapid progress of this fatal species of the disorder. It will prove to them, that the bad effects which have been ascribed to vaccination have been dreadfully exaggerated; and that the temporary duration of its benefits in a few

cases have been owing to some kind of misma-
nagement.

" If such shall be the result of the proposed inquiry, I have no hesitation in saying, that it ought
afterwards to be for this House to consider whether
or not any reward has been bestowed on the original discoverer of vaccine inoculation, which is in
any degree adequate to its real importance; and,
as such, consistent with the general character and
liberality of this country.

" This, however, is a subject for after consideration; but in the meantime the House will agree
with me as to the propriety of collecting opinions
relative to the general effects of this mode of inoculation; and to show to the world that, if there be
any truth as to its benefits, we shall not be the
first to reject them; but that, on the contrary,
we shall use every means to encourage its progress, and this in a manner consistent with the
character and dignity of our nation."

The motion was seconded by Dr. Matthews, in
a judicious and elegant speech. He at that time
sat in Parliament as one of the representatives for
the city of Hereford. The other speakers were
Mr. Wilberforce, Mr. Secretary Windham, Mr.
Bankes, Mr. William Smith, and Mr. Paull. The
sentiments of every speaker were in favour of the
motion; and the whole tone of the debate indicated a state of feeling highly respectful towards
Dr. Jenner, and alive to the value of his disco-

very. Mr. Wilberforce, in his observations, alluded
to a branch of the subject which, though it has
been repeatedly canvassed in Parliament, has not
hitherto met with the consideration and support
that it so evidently demands. His object was not
to force vaccine inoculation, but to impose certain
rules on those who practised small-pox inoculation, in order that the public might be secured
from the effects of that contagion in the same
manner as is done in the case of the plague.
These suggestions were not relished by the House.
Just and moderate though they were, they seemed
to have too much the aspect of compulsion; and
the liberty of doing wrong was still left among the
privileges of free-born Englishmen.

A measure much less comprehensive than that
contemplated by Mr. Wilberforce, should Parliament in its wisdom see fit to enact it, would, even
thus late, be of incalculable service in saving human life, and could scarcely (one would think) be
objected to by the most determined advocate of
liberty. The practice of small-pox inoculation has
been abandoned by almost every respectable medical
man. It has been lately taken up by a set of unprincipled, unfeeling, and ignorant persons. These,
reckless of the miseries which they spread abroad,
extort from the prejudiced parent a pittance sufficient to excite their cupidity; and to show how
small a price, even in this Christian land, is set
upon human health and existence, there is

scarcely a portion of the country where individuals of this stamp may not be traced by the melancholy and fatal consequences of their practices. Were it therefore merely enacted, that no one should practise small-pox inoculation who had not received a testimonial of his qualifications from some of the legally constituted authorities,— this enactment, together with one declaring and enforcing what has been already pronounced by the King's Bench to be the common law of the land, would probably soon banish small-pox inoculation from the kingdom ; but as another occasion will occur of referring to the topic, I shall not dwell on it here.

Sentiments of a similar kind were strongly expressed by some of the most devoted lovers of freedom in this country. His Grace the Duke of Bedford, who then held a highly dignified and responsible public station, that of Lord Lieutenant of Ireland, conveyed to Dr. Jenner a very decided opinion as to the restrictions which ought to be adopted touching small-pox inoculation.

Phœnix Park, Dublin, July 18, 1806.

DEAR SIR,

I am happy to perceive that Lord Henry Petty has introduced the subject of vaccine inoculation once more to the notice of Parliament, though I much doubt whether a bare inquiry into the merits of the practice, or its efficacy against the small-pox, will prove of any essential use unless compulsory means are resorted to, or at least

measures of regulation to prevent the small-pox spreading
its destructive ravages over the empire, and sweeping
thousands annually from the population of the country.
We have already sufficient evidence of the superior and
incontrovertible advantages of the cow pox to satisfy
every rational and candid mind ; and any additional testimony, however sanctioned by the College of Physicians,
or by Parliament, will, I fear, have but little weight in
convincing the obstinate, the interested, or the prejudiced.
Some legislative restraint must be adopted (as I took
occasion to tell you, when I had the pleasure of seeing
you last summer) against that pernicious and fatal error,
which permits a man with impunity to spread the contagion
of a loathsome and cruel disorder around his neighbourhood, and to carry the seeds of disease and death through
the streets of the metropolis, or through towns and villages in the country. This surely cannot be consistent
with the principles of a wise government, or even of a free
one ; and without it, I fear, we shall never effect the great
object the Jennerian Society has in view, the extermination of the small-pox. I have written my sentiments to
Lord Grenville freely on the subject ; and took occasion
to mention my anxious hope, that you would at length
receive that just reward from the public, which in my
opinion has been too long withheld from you. I trust
the inquiry will be extended to Ireland. With the assistance of Dr. Yeates, (whose zeal in the cause you well
know), I am endeavouring to obtain some information on
the progress vaccination has made in this part of the
United Kingdom, which I hope may be useful ; and I am
naturally anxious that Ireland should have her full share
of the benefits resulting from this important discovery.

I should apologise for not having earlier thanked you
for the letter I received from you just before I left Eng-

land. The pressure of public business, and the hurry of
preparations for my departure, prevented my then assuring
you, what I trust you will always believe, that I am,
dear sir,

Your very sincere well-wisher

and faithful servant,

To Dr. Jenner.   BEDFORD.

Though the conduct of the anti-vaccinists was
in every respect unworthy of any notice in Parliament, it is certain that the influence of their
writings tended to produce those results which
rendered parliamentary interposition necessary.
The renowned triumvirate, Drs. Moseley, Rowley,
and Squirrell were quite aware of the effect of
prejudice when duly instilled into the minds of the
fond many; and they availed themselves of every
engine that promised in the least to aid their purpose. Some of their contrivances, indeed, showed
a felicity of invention, which would have been
quite laughable, had they not been altogether
founded in falsehood, and applied to the worst purposes. They actually published prints representing the human visage in the act of transformation,
and assuming that of a cow. There was a *Master
Jowles, the cow-poxed, ox-cheeked, young gentleman,*
and Miss Mary Ann Lewis, the cow-poxed, the cow-
manged young lady, exhibited in all the touching
simplicity of graphic delineation, by Dr. William
Rowley, a learned Member of the University of

Oxford, &c. &c. &c. This august personage, after collecting together the most extraordinary tissue of absurdities, had the folly to characterise his work as " *a solemn appeal, not to the passions of mankind, but to the reason and judgment of all who were capable of deep reflection.*" The voice of the respectable part of the profession was not sufficient to counteract the tide of passion and prejudice which agents, so misguided, had been able to set in motion ; and it is a melancholy view of human nature to be obliged to confess, that the best cause, in the onset, is often foiled by such opponents. They have power enough given them to claim and secure many victims before they are driven from the field ; and it is not till aroused by the magnitude of the evil, that the potency and energy of truth are shown in annihilating the devices of falsehood and error.

Many a pang did Dr. Jenner suffer when he perceived the unhappy success which attended the schemes of the enemies of vaccination. Though the abuse poured out upon himself was most offensive, he regarded it not. A serious reply to such disgusting observations as characterised their productions would indeed have been quite unworthy of him ; but he thought that ridicule was a weapon that might be fairly and effectually wielded against them. In this spirit he actually wrote a letter to one of the chief anti-vaccinists. It was never published; but it contains a great deal of genuine wit and polished irony.

The motion of Lord Henry Petty, "That an address be presented to his Majesty, praying that he will be graciously pleased to direct his Royal College of Physicians to inquire into the state of vaccine inoculation in the United Kingdom, and to report their opinion as to the progress which it has made, and the causes which have retarded its general adoption," was carried unanimously.

The Royal College of Physicians in London having received his Majesty's commands, in compliance with the above address, applied themselves diligently to the business referred to them. They published an advertisement in the newspapers, stating that they were ready to receive information from medical practitioners as to the result of their experience and inquiries upon the subject of vaccination. They had previously applied separately to each of the licentiates of the college. They corresponded with the Colleges of Physicians of Dublin and Edinburgh; and with the Colleges of Surgeons of London, Edinburgh, and Dublin. In short, they took every possible method to gain the most accurate information, whether for or against the practice.

Dr. Jenner himself viewed the whole of these proceedings with the utmost satisfaction. A fair,

a manly, and an unreserved investigation was what
he courted.　He knew that, in such a trial, truth
would come forth in all her clearness and dignity,
and afford the most triumphant refutation of the
calumnies with which he and vaccination had been
assailed.　The inquiry, though it involved the fate
of millions, had this peculiarity, that it was to decide in like manner the character, both moral and
professional, of Jenner himself.　In appearing before the college (which he did on the 19th February 1807) he had to submit to them both testimonies and diplomas granted to him by almost
every learned body in Europe,* in consequence of
the ascertained efficacy of the discovery.　He had
to lay before them grateful tokens of thankfulness
and esteem from the less polished inhabitants of
other quarters of the globe; *all* hailing him as
their greatest benefactor.　Such evidence he could
present, and did present, but he said " trust not to
these, to me cheering and animating documents;
go to my enemies; go to the enemies of vaccination; collect the evidence that they have to offer;
look at their industry in amassing that evidence;
look at the spirit in which it has been put forth;
bring all to the proof; and then let truth prevail."

The college most assiduously and most ably performed the duty committed to them; and, in reporting their opinions on the testimony adduced

* See Appendix No. I. for a list of these honours.

in support of vaccination, they stated that a body of evidence so large, so temperate, and so consistent was, perhaps, never before collected on any medical question. They strongly recommended the practice of vaccination, this conclusion being formed from an irresistible weight of evidence which had been laid before them; adding, that " when the number, the respectability, the disinterestedness, and the extensive experience of its advocates are compared with the feeble and imperfect testimonies of its few opposers; and when it is considered that many who were once adverse to vaccination have been convinced by farther trials, and are now to be ranked among its warmest supporters, the truth seems to be established as firmly as the nature of such a question admits; so that the college of physicians conceive that the public may reasonably look forward with some degree of hope to the time when all opposition shall cease, and the general concurrence of mankind shall at length be able to put an end to the ravages at least, if not to the existence, of small-pox."

On the 29th of July, 1807, the Chancellor of the Exchequer, the late Right Honourable Spencer Perceval, moved the order of the day for the house to go into a committee of supply; and stated that it was referred to that committee to consider of a farther sum to be allowed to Dr. Edward Jenner for the discovery of the vaccine inoculation, and his communication of it to the world. The house resolved itself into a committee accordingly; the late Sir Benjamin Hobhouse being in the chair.

The debate which arose on the motion of the Chancellor of the Exchequer was important; and evinced, with one or two exceptions, a proper estimate of the nature of the question. The excellent report of the College of Physicians afforded the ground on which the Chancellor of the Exchequer proposed "that a sum not exceeding £10,000 be granted to his Majesty, to be paid to Dr. Edward Jenner, as a reward for promulgating his discovery of vaccine inoculation; and that the same be issued without any fee or other reward whatever." The last recited clause was a very considerate one, as in the case of the former grant of £10,000, the fees of office, and other charges, took nearly one-tenth from the sum voted by parliament.

The anti-vaccinists found one advocate (Mr. Shaw Lefevre) in the House of Commons. The honourable member first affected to under-value the merit of vaccination itself; and, next, attempted to show that the discovery of it was not due to Dr. Jenner. Another honourable member* (Mr. Edward Morris) took a very different view of the subject. Influenced entirely by the weight of evidence, he made a powerful appeal to the house; and concluded by saying, "after what I have

Member for Newport, in Cornwall.

seen and heard, and know to have been proved upon this subject, I feel myself called upon to move that instead of £10,000, £20,000 be inserted in this resolution."

Mr. William Smith said, that " every person who would wish to give Dr. Jenner a reward would first allow his expenses; for, until that be done, you cannot talk of reward.   Since, then, the gene- ral merit of the discovery is admitted, and you are about to remunerate the inventor; the first thing you ought to do is to give him back the money which he has been out of pocket in bringing to perfection his discovery—a discovery which has been of so much advantage to mankind."   He then alluded to the little honour which Dr. Jenner had experienced in his own country, and recited part of the valuable information that had just been received regarding the successful termination of the Spanish expedition under Balmis.

After various remarks from other members, the question was put that £20,000 do stand part of the resolution: when the committee divided; ayes, 60; noes, 47; majority, 13.

I have thought it unnecessary to dwell at greater length on the arguments used by the respective speakers on this important occasion. The scientific part of the subject, as well as the character of Jenner, was treated with great elo- quence and effect; but especially by the Marquis of Lansdowne, then Lord Henry Petty.   That noble Lord entered fully into the merits of the discovery, as well as into the disinterestedness and magnanimity of the discoverer. The late Mr· Windham, and the late Mr. Whitbread, spoke likewise with their wonted animation and decision: the former contending that a sum still larger would be more suitable to the character of the country; the latter insisting that Dr. Jenner should be remunerated to the extent proposed by the amendment, the extent of the value of his services being totally out of the question. Other gentlemen alluded to the disgraceful practice which was still carried on of inoculating outpatients with small-pox, at the Small-pox Hospital in London. " I think that the legislature," said Mr. Sturges Bourne, " would be as much justified in taking a measure to prevent this evil by restraint, as a man would be in snatching a firebrand out of the hands of a maniac just as he was going to set fire to a city."

This last subject preyed deeply on the mind of Dr. Jenner. He knew that vaccination would be comparatively powerless while its virulent and contagious antagonist was permitted to walk abroad uncontrolled. In order to restrain this enemy, he sought an audience of the minister. He gives the result in a letter to Dr. Lettsom, dated July 1807.

" You will be sorry to hear the result of my interview with the Minister, Mr. Perceval. I solicited this honour with the sole view of inquiring whether it was the intention of government to give a check to the licentious manner in which small-pox inoculation is at this time conducted in the metropolis. I instanced the mortality it occasioned in language as forcible as I could utter, and showed him clearly that it was the great source from which this pest was disseminated through the country as well as through the town. But, alas! all I said availed nothing; and the speckled monster is still to have the liberty that the Small-pox Hospital, the delusions of Moseley, and the caprices and prejudices of the misguided poor, can possibly give him. I cannot express to you the chagrin and disappointment I felt at this interview."

CHAPTER III.

DOMESTIC LIFE—WORGAN —RETURN OF THE SPANISH VAC-
CINE EXPEDITION—VARIOUS HONOURS—PRESENTS FROM
THE PRESIDENCIES IN INDIA.

REMOTE, though some of these incidents may
appear, from the occupations of a private individual, they made part and parcel of the very existence of Jenner. Many of them may be at this
hour destitute of those qualities which interest
the reader ; but at the time when they occurred
they possessed a power to touch the feelings of
every one ; and were, in an especial degree, important in the sight of Jenner. It is needless,
therefore, to assert how much they engrossed his
attention and occupied his time. In the midst of
such employments we can trace events in his
domestic history which stamp the character of
the man, and bring him before us in a manner
the most engaging and attractive.

His eldest son, Edward, who had a feeble constitution and other infirmities which rendered it

inexpedient to send him to a public school, received all his education under his father's roof.
To assist in this object a domestic tutor was procured. The youth who was selected for this
purpose was but little older than his pupil, but
though tender in years, he was old in wisdom
and knowledge. This extraordinary boy, John
Dawes Worgan, became an inmate of Dr. Jenner's
family at Berkeley in September 1806, having not
then completed his sixteenth year. He was endowed with a singular maturity of judgment, an
uncommon delicacy of perception, a quick and
vivid imagination, a love of high and ennobling
sentiments, together with that deep and heartfelt humility which checked the ardent and impassioned feelings of his nature, and at last
brought all the fond and ambitious imaginings
of his aspiring mind under the sacred influence
of piety and peace.

The feeble texture of his frame and his early
sorrows laid at once open to his mind the real
condition of man, and impressed him in his infancy with those truths which many never learn
till years of bitterness and disappointment have
reduced all earthly objects to their just dimen-
sions. Into the space of a few years he had
crowded the experience of a lifetime; so much so,
that in reading some of his " Remains," the
strength and the confidence of an aged and well-
taught pilgrim may be discerned, rather than the

hesitating and crude conceptions that for the
most part characterise the productions of a youth
of seventeen. I speak more especially of his
sentiments on those questions which the human
mind does not always receive with favour. His
parents were distinguished by their Christian faith
and practice, and their great object was to instruct him how to live and how to die. These
purposes were subsequently much promoted by
his residence at a school in the village of Ful-
neck, near Leeds, under the direction of the
United Brethren. His parents being in humble
circumstances, and unable to bestow upon him a
learned education, he was designed for some mercantile employment, and he actually assisted his
father in his occupation of a watchmaker. After
the death of this parent, in 1803, he announced
to his mother a design, which he had long entertained, of dedicating himself to the service of God,
as a minister of the Church of England. The
excellent and venerable the Rev. T. T. Biddulph,
of Bristol, who had been his friend and adviser,
both in temporal and spiritual matters, and who,
to the most unreserved sincerity, had joined the
most tender and affectionate kindness, was consulted on this occasion. The direction of his
studies was in consequence immediately changed.
He now began to devote himself to the learned
languages; and the felicity of his genius was such,
that he was able in a very short time to master all

difficulties, to pass through the steps that lead
to sound and good scholarship, and to shew at
the same time a fidelity of memory, an expansion
of intellect, and a delicacy and correctness of
taste, that were rarely to be met with. These
qualities brought him into the family of Dr. Jenner : here his genius was cherished, his sensitive
nature was protected from many evils to which
the hardness of his fortune and the roughness of
the world would otherwise have exposed him.
Jenner loved to hold converse with such beings,
and the whole atmosphere of his domestic circle
was at that time in unison with Worgan's keen
and deep-toned feelings.   Mrs. Jenner was in
very delicate health, and her soul was devoted
to those contemplations which most delighted
Worgan.

During the year in which he became an inmate
in Dr. Jenner's house, the Spanish vaccine expedition, under Balmis, had returned to Europe : his
correspondent, the late Mr. Hayley, had suggested
this as a fit subject for the muse of Worgan, who
almost from his infancy had shewed a decided
taste for metrical composition.   He himself had
often projected such a work, but at that time he
did not feel that his wing was strong enough to
soar into the regions of historic verse.   He wrote,
however, an address to the Royal Jennerian Society, which was printed and presented to the
members at their annual meeting, in 1808.

He had two attacks of typhus fever while under Dr. Jenner's roof, and he was finally destroyed by pulmonary consumption in 1809. The unceasing energy of his mind could ill bear the hours of listlessness and inactivity which the feebleness of his frame needed; the intensity of his feelings, and the eagerness of his nature, hurried him on to exertions which he could not sustain. The flame was burning with too great vehemence, and it became extinguished ere it had acquired all its brilliancy and strength.

Dr. Jenner's own sentiments may be gathered from the following extract of a letter which he wrote to a friend immediately after Worgan's death.

" Your letter of course came too late for me to make any observations upon what you drew up for insertion in the Bristol paper, respecting our dear departed friend. It must be some consolation to his surviving relations and friends that his name will not be forgotten, and greater still to those that were most dear to him, that his long indisposition awakened in him those sentiments, in all their purity, from which alone can spring true happiness at any period of our existence, but particularly at ' the awful hour of death.'

" I beg you will present my best wishes to Mrs. Worgan, and tell her, unless she particularly wishes it, I should be sorry to put her to the expense of a ring; but yet, I should like to have something in remembrance of poor John. A book would be acceptable. The editor of the Chelten-ham Chronicle (Mr. Pruen), who was the intimate acquaintance of poor Worgan, has paid a just tribute to his memory, in the paper of yester-day."*

* The following poetic tribute to the memory of this amiable, accomplished, and pious young man, is the heart-felt effusion of one of his intimate associates in friendly and literary intercourse; and who, though Worgan's senior by several years, has candidly acknowledged, that whatever refinement and polish his slender vein of poetic ore has attained, he owes to his young friend's superior judgment and taste.

To the memory of John Dawes Worgan, who died on the 25th of July, 1809, aged 19 years.

While JENNER'S fost'ring hand was stretch'd to save
Thy genius, Worgan, from th' untimely grave,—
While the fond Muse thy wit and fancy shared,
And for thy brow an early wreath prepared,—
Heav'n claim'd *thy heart*;—and, to assert the claim,
Snatch'd thee from dang'rous paths of earthly fame;
Then gave thee—rich exchange for such renown!
Immortal bliss, and a celestial crown!

                                        J. B. DRAYTON.†
Cheltenham, 1809.

---

† Mr. D. had the privilege of being on terms of friendly intercourse and correspondence with Dr. Jenner, and having been trained as a professional artist, under T. Holloway, Esq. the celebrated engraver of the Cartoons of Raphael, he, in the

The philanthropic expedition under Balmis has been already more than once alluded to. The tidings of its safe return to Spain reached this country in the autumn of 1806. A copy of the Madrid Gazette, which announced that important event, was immediately forwarded to Dr. Jenner. This document, recording one of the most remarkable incidents in the history of vaccination, was kindly translated by the Marquis of Lansdowne. In conveying the translation to Dr. Jenner, his lordship accompanied it with the following gratifying letter.

THE MARQUIS OF LANSDOWNE TO DR. JENNER.

Cheltenham, Nov. 18th, 1806.

DEAR SIR,

I send you a translation of the official account * of the vaccine expedition, undertaken by command of his Catholic Majesty, which will, I hope, be found to possess the merit of fidelity. The importance of your discovery

year 1805, solicited and obtained for his own gratification, and by way of complimentary present to Dr. Jenner, a sitting for his portrait (a small medallion profile in pencil, with a view to assist in forming a die for a gold medal), which is allowed to be one of the most exact resemblances of Dr. Jenner, and was engraved by the late Mr. Anker Smith, at the Doctor's expense, for private circulation. The engraving is a pleasing specimen of the art.

* See Appendix, No. II.

will be much better comprehended by those who have been in the habit of occupying or frequenting countries characterized by heat of climate, than by those who have constantly enjoyed the advantages which belong to a temperate region. You have conquered more in the field of science, than Buonaparte has conquered in the field of battle; and I sincerely congratulate you on so glorious a testimony of your success, as that which the Spanish narrative affords.

I am,

Dear Sir,

Yours very sincerely,

LANSDOWNE.

The important narration alluded to in his lordship's letter stated, that on Sunday the 7th of September, 1806, Dr. Francis Xavier Balmis, Surgeon-Extraordinary to the King of Spain, had the honour of kissing his Majesty's hand, on the occasion of his return from a voyage round the world, executed for the sole object of carrying to all the possessions of the King of Spain beyond the seas, and to those of other nations, the inestimable gift of vaccine inoculation. The reader will recollect that the expedition, of which Don Balmis was the director, sailed from Cadiz on the 30th of November, 1803. It made the Canary Islands first; it then proceeded to Porto Rico and the Caraccas. On leaving the port of La Guira, it was divided into two branches, one part sailing to South America, under the charge of the sub-director,

Don Francis Salvani; the other, with Balmis on
board, steering for the Havannah, and thence for
Yucatan. There a sub-division took place. The
professor, Francis Pastor, proceeded from the
port Siral, to that of Villa Hermosa, in the province of Tobasca. The rest of the expedition
traversed the vice-royalty of New Spain and the
interior provinces; and thence returned to
Mexico, the point of re-union. This being accomplished, the next object of the director was
to carry the preservative from America to Asia.
After surmounting various difficulties, he em-
barked in the port of Acapulco for the Philippine
Islands, carrying with him from New Spain twenty-
six children destined to be vaccinated in succession. The cow-pox having thus been disseminated
through the islands subject to his Catholic Majesty, it was originally designed that the expedition
should then terminate. The director, however,
and the Captain-General, concerted the means of
extending the beneficence of the King to the re-
motest confines of Asia. Setting sail, therefore,
for Macao and Canton, they introduced the preservative to the Portuguese settlements, and to
the inhabitants of the vast empire of China. Balmis returned from Canton to Macao, and em-
barking in a Portuguese vessel reached Lisbon on
the 15th of August, 1806. In his way he touched
at St. Helena; and, strange to say, was the first
to induce the English inhabitants of that settle-

ment to adopt the antidote ; and this even, though it had been discovered in their own country, and sent to them by Jenner himself.

The fate of that part of the expedition destined for Peru was disastrous, having suffered shipwreck in one of the mouths of the river La Magdalena. Providentially, the sub-director, the members of the faculty, and the children, with the fluid in good preservation, were saved. It was thence carried to the Isthmus of Panama. Another part of the expedition ascended the river La Magdalena. When they reached the interior, they separated, to discharge their commission in the towns of Teneriffe, Mompox, Ocana, Socorro, San Gil y Medellin, in the valley of Cucuta, and in the cities of Pamplona, Giron, Tunja, Velez, and other places in the neighbourhood, and reunited at Santa Fe. Towards the close of 1805, they again separated, for the purpose of traversing the remaining districts of the vice-royalty, passing by Popayan, Cuença, and Quito, as far as Lima. In the August following, they reached Guaiaquil.

Not one of the least remarkable events in this expedition was the discovery of the indigenous cow-pox in three different places; namely, in the valley of Atlixco, in the neighbourhood of Valladolid de Mechoacan, and in the district of Calabozo, in the province of Caracca.

The conductors of the expedition were everywhere welcomed with the utmost enthusiasm. It

was to be expected that the representatives of the
Spanish monarch and all the constituted authorities would gladly co-operate; but it was scarcely
to be anticipated that the unenlightened minds of
the Indians would so soon appreciate the value of
the mission. It is, nevertheless, most gratifying
to know, that the numerous hordes which occupy the immense tract of country between the
United States and the Spanish colonies, all received the precious fluid with the utmost readiness. They acquired the art of vaccinating, and
soon performed the operation with great dexterity.

Fame had preceded the arrival of Salvani at
Santa Fe. On approaching the capital he was
met by the viceroy, the archbishop, and all the
civil and ecclesiastical authorities. The event was
celebrated with religious pomp and ceremonies;
and in a short time more than fifty thousand persons were vaccinated. Similar honours awaited
the expedition throughout its whole course. At
Quito they were greeted with boundless joy and
festivity. Such expressions well became them.
The people of this country, the Indians more especially, having been often scourged by the horrid ravages of small-pox, regarded it as the
most terrible affliction which Heaven could send
them. On its first appearance in a village a panic
seized every heart; each family prepared an
isolated hovel, to which those who were supposed
to be infected were banished. *There,* without suc-

cour, without remedy, and with a very insufficient supply of food, they were exposed to the alternations of a very variable climate, and left to their fate. In this way whole generations perished. Under the viceroy Toledo the population of the native Indians had amounted to seven millions and a half; at the time of this expedition the number was supposed to be reduced to one-fifth.

I fear that the energy which prompted the measures on the first introduction of vaccination into South America has not been maintained : I perceive, at least, by a message from the government of Buenos Ayres to the seventh legislative assembly, dated September 12th, 1828, that small-pox had been making dreadful ravages in the adjoining districts. The words of that part of the message relating to this topic deserve to be recorded in this place. "The important establishment of vaccination has been augmented, and its utility has never been more felt than at this moment : whilst the neighbouring provinces are visited by the terrible scourge of the small-pox, it has scarcely been felt in this city, and the government has put in practice every means entirely to eradicate it."

The arrival of the Spanish expedition at Macao was followed by very extensive vaccination at that place, as will appear from the following extract from a letter, dated Macao, 28th of June, 1805 :—

"Mr. Pearson devotes one day in the week to vaccine

inoculation, and the numbers brought to him rapidly increase. Two of our compradores, and Mr. Drummond's head servant, have been taught the mode of vaccinating, and have daily applicants from the neighbouring villages. The treatise drawn up by Mr. Pearson, and translated by the compradore and Sir George Staunton, has been sent to Canton, and 200 copies will be immediately struck off. Advertisements have been also sent to Canton, directing those who are desirous of promoting this blessing, to bring the subjects for inoculation to the company's factory, where it will be performed gratis, and the medical men be supplied with lancets ready charged, and copies of the above treatise. In the title-page are mentioned Mr. Pearson, the English surgeon, and Sir George Staunton, who accompanied Lord Macartney to Pekin, and now in the company's civil employ; and Gnewgna, one of the Hong merchants, being here on a visit, Mr. Drummond prevailed on him to sanction the publication by adding his signature.

" This intelligence will be highly grateful to Dr. Jenner, as well as to humanity at large, when it is known that *one-third* of the people of this extensive empire, when the natural small-pox is raging, are supposed to fall victims to it.

" P.S. I cannot close this without informing you that the Jennerian system of inoculation has been introduced into this place by means of subjects from Manilla, and many children, both Portuguese and Chinese, have been vaccinated with complete success; and am happy to add that such of the Chinese as have witnessed the innocence of the operation, and the mildness of its effects, are now sensible of the importance of the discovery, and I have no doubt of its being shortly practised throughout the empire.

G 2

"Every thing is now arranged for the departure for Pekin of the two French missionaries who came out in the Dorsetshire, to whom Mr. Pearson has forwarded large supplies of lancets charged with the genuine virus, in the view to the introduction of this inestimable blessing throughout the empire, accompanied with a short plain treatise on the subject by himself, and translated into the language of the country by Sir George Staunton; which is certainly a very humane and praiseworthy undertaking."

In the course of a few months Dr. Jenner had the satisfaction of receiving still more gratifying intelligence from this quarter. The circumstances to which I refer were made known to him by a distinguished individual well acquainted with the character of the Chinese.

MR. (NOW SIR JOHN) BARROW TO DR. JENNER.

3, *Charles-street, St. James's Square,*
9th June, 1806.

SIR,

I have great pleasure in being able to inclose for your inspection a short treatise in the Chinese language on the vaccine inoculation, translated by my friend Sir George Staunton, and published by the Chinese in the city of Canton. The curiosity of an English work issuing from the Chinese press, however extraordinary, gives way to the more extraordinary facility with which this people, always strenuous in opposing every innovation, has submitted to receive the new practice of vaccination. Not only the surgeon of the English factory, but numbers of the Chinese were constantly employed in communicating the disease from the moment it was perceived with what ease and convenience the patient went through it; and they had actually raised a very considerable subscription for the purpose of establishing a vaccine institution for promoting the practice in every part of this extensive empire. Thus the English, at length, as well as the other Europeans, have established their claim, which, though last, is not the least, on the gratitude of the Chinese.

As the small-pox in China has usually been attended with most fatal effects, there is little doubt that the same willingness, which has manifested itself at Canton, to receive so mild and effectual a substitute, will be felt in every province of this populous country ; and the more so, as public confidence there is not likely to be shaken by that kind of illiberal and undignified opposition which has been so industriously employed elsewhere.

By every real friend of humanity, and by you, sir, in particular, this intelligence must be received with sensations of peculiar satisfaction.

I have the honour to be, Sir,

Your most obedient and humble servant,

JOHN BARROW.

P. S. As I imagine the inclosed to be the only copy in England, you will be pleased, at your convenience, and when done with, to put it under cover to me."

The heartfelt satisfaction which Jenner experienced on the successful issue of the Spanish expedition in some measure repaid him for the anxiety and suffering which, up to that time, he had endured from the unjust and ungenerous attacks made upon him by too many of his own countrymen. He transmitted copies of the Madrid Gazette to all his friends, and only regretted that the honour of the beneficent enterprise which it recorded should not have been acquired by England. His Grace the Duke of Bedford, in acknowledging the receipt of one of those documents, thus writes:—

THE DUKE OF BEDFORD TO DR. JENNER.

Dublin Castle, Dec. 9, 1806.

MY DEAR SIR,

By yesterday's mail I was favoured with your letter of the 4th inst. and am much gratified by the accounts you send me of the progress of vaccination in distant quarters of the globe, although I cannot but lament with you that Great Britain has suffered other nations to take a more conspicuous part than herself in the extension of this inestimable blessing.

You will readily believe that the importance I have ever attached to the success of the vaccine system induced me to direct my attention to this subject at an early period of my government in this country, and I am persuaded that you will be satisfied with the Report which has been drawn up by Dr. Yeats, and which I have now transmitted to the College of Physicians.

Dr. Yeats is at present in England for a short time; but on his return, I will not fail to give him the very satisfactory account from the Madrid Gazette. He added to his Report the outline of a plan for extending vaccination and exterminating the small-pox throughout the British dominions (and indeed throughout the world); but as this did not form any part of Lord Henry Petty's motion, I did not send it to the College; but, if you should desire to see it, I will with pleasure forward you a copy of this paper.

I remain, with very sincere regard,

My dear Sir,

Your faithful and obedient,

. BEDFORD.

At this period more substantial proofs of consideration reached Dr. Jenner from our Indian possessions. The idea, which was started by Dr. Lettsom in England, had about the same time occurred to several of the European inhabitants in the different presidencies. They almost simultaneously resolved to present him with a testimony of their gratitude. An announcement of their proceedings, together with Dr. Jenner's reply, is subjoined.

To DR. EDWARD JENNER, BERKELEY.

Calcutta, 17th May, 1806.

SIR,

The principal inhabitants of Calcutta and its dependencies having some time ago resolved to present you with a testimonial of their gratitude for the benefit which this settlement, in common with the rest of mankind, has derived from your inestimable discovery of a preventive of the small-pox, and having appointed us a Committee for carrying their resolution into effect, it is with the highest satisfaction that we now discharge the duty committed to us, by transmitting to you herewith bills drawn on the Honourable Court of Directors to the amount of three thousand pounds sterling. Duplicates and triplicates of these bills, together with the remainder of the subscription (about one thousand pounds), will be hereafter forwarded to you by the first favourable opportunity.

We have the honour to be, with the greatest esteem and regard,

Sir,

Your most obedient,

humble servants,

ROBERT SMITH,   J. FLEMING,

H. COLEBROOKE,   J. ALEXANDER.

Per Charger.

DOCTOR JENNER TO DOCTOR FLEMING.

MY DEAR SIR,

The death of my great and good friend Lord Cornwallis, and the uncertainty whether my letter, directed to you at Portsmouth when you were upon the eve of sailing, ever reached you, were circumstances which nearly set aside every hope of my being brought forward at Calcutta as one deserving public attention. But your unexpected letter of 25th March, told me my fears were groundless. Since that time, I have the pleasure of informing you I have duly received the bills remitted, and the very handsome letter which accompanied them. In my inclosed reply to this letter, you participate only in my thanks with my three other friends, Mr. Smith, Mr. Colebrooke,

and Mr. Alexander, who kindly laboured with you in my
cause. But now you must allow me to express the very
particular obligations I feel under to you. Indeed, it is
highly probable that without your industry and benevolence, the good people of Calcutta, like most other parts of
the world, would not have expressed their thanks in the
way they have done. Permit me to request you to
present my particular thanks to the medical gentlemen you name to me, who have been so strenuous in
promoting the donation. Mr. Russell's acquaintance
I had the pleasure of making at Cheltenham and re-
newing in town. Mr. Hare I have not the pleasure
of knowing, but from his high reputation. Mr. Shoolbred has been my correspondent, and has my warmest regards for his excellent publication on vaccination. Some
disaster, I fear, has befallen my late dispatches to him, as
I have received no answer to my letters.

No intelligence of any sort has of late reached me from
Madras or Bombay; but I once heard that, at the latter
place, the inhabitants intended me a pecuniary compliment.
Vaccination, I find, has been unbounded in both these
settlements.

I have sent you some copies of a Madrid Gazette.
What a glorious enterprise! Yet, while I feel proud in
contemplating it, I cannot but lament that it was not
achieved by the British nation. To say the truth, this country has been dreadfully supine in the matter hitherto: however, its energies seem roused, as you will see by the inclosed
advertisement. Some pamphlets, full of the grossest
misrepresentations and forgeries, have been spread ; and
the common people became so terrified, particularly
when told that their children, if vaccinated, would take the
similitude of bulls and cows, that a great dislike to the
practice has arisen among them : and these accounts have

been circulated through the country with peculiar industry. The consequence has been the re-introduction of variolous inoculation, which has produced an epidemic small-pox through the metropolis and the whole island, except in those parts where vaccination had previously been so generally adopted as to forbid its approach. This, now too late, has opened their eyes, and they see the powers of the cow-pox. The folly of the oppositionists has gone so far as to exhibit prints of children undergoing transformation from the human being into that of a brute.

Believe me, dear Sir,

With the warmest sentiments of esteem,

Your obliged and very faithful servant,

EDWARD JENNER.

In a very short time after this period, an authentic report of what was doing in Bombay reached him. A letter from Dr. Helenus Scott conveys *this* and other interesting intelligence.

DR. HELENUS SCOTT TO DR. JENNER.

Bombay, 5th December, 1806.

MY DEAR SIR,

I very lately received your letter of the 8th February last, with the evidence on the vaccine subject, and the Report, &c. It is certainly not to Bombay that you need take the trouble to send further evidence. We have long had abundance of it, and daily experience gives us more and more. The attacks on vaccination have made but very little impression here, none whatever on myself, or on those medical men for whose judgment I have any value. I have given Mr. Duncan the report you sent for him, and I have shewn him the letter you wrote me, which contains many obliging expressions with respect to himself. He desires me to thank you, and to assure you of his esteem and high regard.

Some centuries ago men were too credulous. The limits of many branches of knowledge were less known than they now are, and they had no kind of rules by which belief could be limited. We have now gone to the contrary extreme, no less unphilosophical, and perhaps more injurious to science, of crediting nothing that is new and unexpected. Of all the circumstances that have occurred to me, the resistance that has been made to the cow-pox in England was what I expected the least. In all that you have asserted, so much is evident, and, I may say, palpable, that I looked for no resistance.

* * * * *

That all this should have been rejected (I must still affirm, without sufficient inquiry), I have never been surprised; but that the cow-pox should stop in its progress, has filled me with wonder. It is so interesting to every individual, peculiarly so to every parent, that I thought (against the practice of this world) the triumph was complete, even in the lifetime of its inventor.

I confess that I am much displeased with what I observe at a distance of many members of our profession. They dispute with violence and ill manners ; and on many occasions one is tempted to believe that he has gotten into very bad company. I suppose that those who are the most disposed to fight and scribble, are not in much estimation by such as know them better than we can do at this distance. I frequently feel concern for the profession.

It is now several years since Dr. Duncan, several other

gentlemen, and myself, made a proposal that some testimony of our thanks should be offered to the discoverer of vaccine inoculation. A paper was circulated to that effect, which from various causes has been long kept from the Presidency. You must know that our territory on this side of India, and consequently the British inhabitants, are now scattered over a vast tract of country. This paper has been at Sazurat, and with our army I believe at Delhi. Although our power extends over a great space, the number of British inhabitants is not very considerable. From these and other causes, this testimony of our thanks has been long delayed, and will at last I fear be less considerable than we could wish. It will, however, shew our intentions, and I hope on that account be agreeable to you. When all is settled, I intend that the Medical Board, (of which I have the honour to be President), shall bring the matter before government, and beg of them to remit the amount of the subscription to the Court of Directors.

The vaccine inoculation goes on here with its usual success. In this island, swarming with mankind, no loss has been suffered by the small-pox for several years, since the introduction of the vaccine inoculation. I shall desire the Secretary to the Medical Board to transmit to you the monthly reports of vaccination, which are published by order of this government.* ”

Vaccination having been happily established in all the governments of European Russia, his Ma-

---

* The full amount of the contributions from our possessions in the East was not received by Dr. Jenner till 1812. From Bengal was remitted £4000 ; from Bombay £2000 ; from Madras £1383. 1s 10d.

jesty the Emperor commanded Dr. Boutlatz' to traverse the remaining parts of his vast empire, and to spread the practice in every direction. This gentleman had studied in England. He did not visit the two populous districts of Nisny-Novogorod and Casan. In the former, the physicians resident in the families of the nobility interested themselves in propagating this mild disease. The Minister of State had his officers instructed in the manner of conducting this simple operation; and himself vaccinated several children in the villages. The Prince of Georgia actively promoted the practice among the children of his dependents, and of all who resided in his vicinity. At Casan the practice had been introduced before the Mission reached that town, by M. Walkoff. His exertions succeeded in disposing the Tartar merchants to adopt the new practice. He caused to be translated a work on vaccination compiled by the Medico-Philanthropic Society of Petersburgh. This translation was printed at the Tartar press established at Casan.

The surgeon Stury found that it had made its way into Siberia; Dr. Grahl, at Perm, had been indefatigable in spreading it. He had vaccinated in two years six hundred persons. From him the mission received the last supply of fresh matter; and from his stock all the other districts of Siberia were supplied. The mission proceeded to Tomsk, then to Krasnojoesk. They next advanced to the capital of Siberia, and lastly to Jakoutzk and Ochotzk. From thence the virus was transmitted to Kamstchatka, and the islands situated between Asia and America, extending thus to the north-eastern extremity of our hemisphere.

At Irkoutzk, where the Empress Katherine had founded an institution for small-pox inoculation, but which had been disused for many years, several children were vaccinated, in order that a fresh and certain supply of virus might be had for transmission to China. The Russian and the Spanish expeditions thus reached different points of the celestial empire nearly at the same period of time.

In the United States of America vaccination continued to maintain its high character. The President Jefferson, whose early service in that cause has been commemorated, evinced his feelings in the following emphatic manner.

MR. JEFFERSON TO DR. JENNER.

Monticello, Virginia, May 14, 1806.

SIR,

I have received the copy of the evidence at large respecting the discovery of the vaccine inoculation, which you have been pleased to send me, and for which I return you my thanks. Having been among the early converts in this part of the globe to its efficacy, I took an early part in recommending it to my countrymen. I avail myself of this occasion to render you my portion of the tribute of gratitude due to you from the whole human family. Medicine has never before produced any single improvement of such utility. Harvey's discovery of the circulation of the blood was a beautiful addition to our knowledge of the ancient economy; but on a review of the practice of medicine before and since that epoch, I do not see any great amelioration which has been derived from that discovery. You have erased from the calendar of human afflictions one of its greatest. Yours is the comfortable reflection that mankind can never forget that you have lived; future nations will know by history only that the loathsome small-pox has existed, and by you has been extirpated. Accept the most fervent wishes for your health and happiness, and assurances of the greatest respect and consideration.

Th. Jefferson.

The Royal College of Physicians of Edinburgh, on the 20th of May 1806, unanimously elected Jenner an honorary fellow of their college. Dr. Spens, the President, in announcing this distinction, expressed himself in the highest terms of respect for the eminent individual whose great merit he was then acknowledging.

The anniversary of the Jennerian Society, as usual, was held on the 17th of May 1807; his Royal Highness the Duke of York in the chair. Though Jenner felt a peculiar dislike to exhibitions of this kind, he seems to have been much gratified with the events of this day. Dr. Lettsom and the Rev. Rowland Hill energetically

supported vaccination. Jenner, too, on his own
health being drank, spoke with much delicacy and
propriety. " After the very animated speech of
the Duke of York, the illustrious chairman, and
the important information conveyed by his friend
Dr. Lettsom, he had little to say on the subject.
He continued to receive the most agreeable information respecting vaccination from all parts of
the world, from Greenland to the Cape, and from
the Mississipi to the Ganges." He then alluded
to the effects of prejudice and falsehood in retarding the progress of vaccination in the metropolis;
and he refuted a calumny which had been put
forth to the public respecting failures alleged to
have taken place in his own practice.

At this meeting, an account of the introduction
of vaccination into China was delivered by Mr.
Parry, one of the East India Directors. Mr. Ring
read a translation of a Latin letter from Dr. Reyss,
of Makow in Poland. It was addressed to the
" Illustrious exterminator of that pestilential disorder the small-pox." He complimented Jenner
highly on his discovery; sent him a richly em-
bossed silver cup, which had belonged to a person of the name of JENNER; wished that joy and
festivity might prevail on his birthday; and requested to be enrolled among the honorary members of the society. He requested also a portrait
of Jenner, and a small pattern of the cloth that he
generally wore, that he and his friends might

wear the same garb on the 17th of May, the birthday of the DISCOVERER of VACCINATION. In returning thanks to Dr. Reyss for his letter and his present, Jenner observed, " I may with truth say, that greater attention was never paid me by any individual since the first promulgation of the vaccine discovery. My native county has complimented me with some very elegant plate; but there is not a piece among it I set so high a value upon as the curious antique cup you have presented to me, and which has probably graced the table of those who sprang from the same stock as myself."

Writing to Mr. Ring, concerning the hearty and animating letter of Reyss, Jenner observes, " John Reyss is a fine fellow. I have given order to Manning to prepare a bust for him, thinking it would be more acceptable than the print."

This bust, executed by Manning, is indeed an extremely faithful and valuable likeness. Some years after the period just mentioned, he presented me with one, and I value it as a most satisfactory resemblance, not less than for the donor's sake.

CHAPTER IV.

GENERAL VIEW OF THE STATE OF VACCINATION—ADDRESS
OF THE FIVE INDIAN NATIONS— INFLUENCE OF JENNER
WITH FOREIGN STATES— FORMATION OF THE NATIONAL
VACCINE ESTABLISHMENT—CAUSES OF HIS WITHDRAWING
FROM IT—DEATH OF HIS ELDEST SON.

WE are now arrived at a period when it may be
judicious to take a very brief survey of the state of
vaccination throughout the world. Had I been
writing the history of the practice, it would have
been necessary to have traced its progress with
greater minuteness. It had been eagerly espoused
by the King of Prussia, the Emperor of Germany,
Napoleon, the Emperor of Russia, the King of
Spain, the King of Naples, the King of Denmark,
the King of Sweden, the Elector of Swabia, the
Queen of Etruria. It had gained the sanction of
the Grand Seignior, the Dey of Algiers, the Hos-
podar of Moldavia, the heads of our government
in India, many of the native chiefs, and, under the
authority of the governments, had been diffused

through all the states both in North and South America. It had taken firm root in Siberia and China; and had been carried to many of the islands in the Pacific and Indian oceans, and to most of our possessions in the West Indies.

Of course it is impossible to ascertain with accuracy the number that may have been vaccinated up to this period; but there is good reason for believing, that it must have been considerably more than twenty millions. Indeed, when we reflect that Dr. Sacco and his assistants vaccinated, in eight years, thirteen hundred thousand persons, I am satisfied that I am giving a very low estimate.

Besides this very large number vaccinated by professional gentlemen in different parts of the world, it is but proper to commemorate the services of many ladies and gentlemen in England, who particularly distinguished themselves by their efforts in this cause. Among them ought to be mentioned the Rev. Rowland Hill *, W. Bramston

* On the discovery of vaccination by Dr. Jenner, Mr. Rowland Hill eagerly embraced this new means of conferring a benefit on his fellow creatures, and ably defended it against its opponents. " This," he said, " is the very thing for me; " and wherever he went to preach, he announced after his sermon, " I am ready to vaccinate to-morrow morning as many children as you choose; and if you wish them to escape that horrid disease the small-pox, you will bring them." Once a week he inoculated the children who were brought to him from Wotton and the neighbourhood; and it is well known that one of the most effective vaccine boards

of Oakeley Hall, Hants, Esq. the Lady Charlotte
Wrottesley, Miss Bayley of Hope, near Manchester, Miss Cox of Painswick, in the county of Gloucester, Mrs. Kingscote of Hinton House, Hants,
T. Westfaling of Rudhall, near Ross, Hereford-
shire, Esq. the Rev. W. Finch of St. Helen's, Lan-
cashire, the Rev. J. T. A. Reed, Leckhampstead,
&c. &c. I have already stated, notwithstanding

in London was established, and still continues in operation,
at Surrey Chapel.

When vaccinating the children, he seemed quite in his
element, talking kindly to the parents, and coaxing the little
frightened creatures in the most good natured manner. In
a few years the numbers inoculated by him amounted to
more than ten thousand; and in most of the cases he was
particularly successful. Dr. Jenner was of a very lively turn
of mind * and animated conversation, with a remarkably kind
disposition; and although he did not fully participate in his
venerable friend's religious views and feelings, he had the
highest reverence for his character, and was a frequent attendant on his ministry at Cheltenham. He seemed at
times forcibly struck with the deep tone of the zealous
preacher's piety and glowing anticipations of happiness in a
spiritual state of being. Mr. Hill once introduced him to a
nobleman in these terms: "Allow me to present to your
Lordship my friend Dr. Jenner, who has been the means of
saving more lives than any other man."—"Ah! would I,
like you, could say *souls*."—Sidney's Life of the Rev. Rowland Hill, pp. 225, 226.

---

* I remember seeing these two remarkable men amusing
themselves in playing with an old eagle in Dr. Jenner s garden at Berkeley, with all the sportive interest of boys.

the clamour which was raised against unprofessional vaccinators, that they rendered good service to the cause. They were obedient and teachable, and certainly in the main conducted the practice with greater success than many professional persons. I have often heard Dr. Jenner speak with satisfaction of the conduct of the ladies. Miss Bayley in particular managed the process with much skill and perseverance. In order to detect any cases of failure that might occur in her practice, she told the poor that she would give a reward of five shillings to any one who could produce an instance of small-pox after vaccination performed by her. Out of 2600 cases, only one was brought to claim the reward; but on referring to her journal, she found a mark against the name, indicating her belief that vaccination had not properly taken effect.

Mr. Kingscote's vaccinations were nearly as extensive, and, I believe, quite as perfect as those of Miss Bayley.

After various attempts to disseminate vaccination among the native tribes of North America, Dr. Jenner at length had the happiness of finding that his efforts were successful. He had sent through the hands of Colonel Francis Gore, Lieutenant-Governor of Upper Canada, his work on vaccination, to be presented to the Five Nations. On the 8th of November, 1807, they assembled in

Council at Fort George, in Upper Canada, to receive this gift, and to reply to Dr. Jenner. They
were addressed by William Claus, Esq. Deputy
Superintendent-General of Indian affairs, as follows :—

BROTHERS OF THE FIVE NATIONS,

Early in May last, His Excellency Lieutenant-Governor
Gore took every possible means to introduce vaccine inoculation among your tribes; but, owing to your people
being then out on their hunt, it did not take place. When
on public business here about a month after, I spoke to
you again, and strongly recommended to your serious
consideration the introducing among your people this
valuable discovery, the want of which you soon afterwards felt very severely in the loss of one of your chiefs,
Oughquaghga John.

Brothers! I have now the satisfaction to deliver to
you a book, sent to you from England, by that great man,
Dr. Jenner, whom God enabled to discover so great a
blessing to mankind: it explains fully all the advantages
derived from so great a discovery.

I, therefore, Brothers, at his request, and in his name,
present this book to the Five Nations, as a token of his
regard for you and your rising generation, by which many
valuable lives may be preserved from that most dreadful
pestilence, the small-pox.

(Signed)   W. CLAUS, D.S.G.I.A.

*Speech of the Five Nations, assembled in Council at Fort
George, in Upper Canada, to Dr. Jenner, London, on
the 8th of November, 1807.*

Present,

Lieutenant Colonel Proctor, 41st regiment, commanding the garrison.

William Claus, Esq. Deputy Superintendent-General of
Indian affairs.

Lieutenant Saunders, 41st regiment.

Ensign Bullock, 41st regiment.

Lieutenant Fowler, 41st regiment.

W. J. Chew, Storekeeper, Indian Deputy.

David Price,   } Interpreters.
Benjamin Fairchild,

Brother! Our Father has delivered to us the book
you sent to instruct us how to use the discovery which
the Great Spirit made to you, whereby the small-pox,
that fatal enemy of our tribes, may be driven from the
earth. We have deposited your book in the hands of the
man of skill whom our great Father employs to attend us
when sick or wounded.

We shall not fail to teach our children to speak the
name of Jenner; and to thank the Great Spirit for bestowing upon him so much wisdom and so much benevolence.

We send with this a belt and string of Wampum,* in

* The Wampum is at once the current coin of the untu-
tored Indian, and the emblem and the pledge of all his con-
tracts. It ratifies his private friendships, as well as the
most solemn and important public treaties. A string of
Wampum passed from one hand to another is sufficient for

token of our acceptance of your precious gift; and we be-
seech the Great Spirit to take care of you in this world
and in the land of spirits.

Chiefs' Names.	Tribes' Signatures.	Interpretation of the Names.	Nations.
Dewataharanegea		Two pointed arrows .	} Mohawks.
Dekayonwagegh		Two Wampum Belts.	
Aigowane . .		Clear Sky . .	} Onondaga
Auneai . .		Feathers on his head ·	
Cosscouete . .		Moving a tree with brush, and planting it . . .	} Senecas.
Onindaki .		A Town Destroyer .	
Caugheaw .		Raven . . .	} Oneidas.
Ussweghtagehte .		Belt Carrier . .	
Sawesyewathaw .		Disturber of Sleep .	} Cayougas.
Ejaahtewge .		Fish Carrier . .	

the former of these purposes. Strings multiplied and united
together in the form of a belt, varying in length and breadth
according to the importance of the occasion, bespeak treaties
of a different description. Jenner's belt, from its size, I
should imagine must have been such as they used for these
latter purposes.

That such tokens and assurances of regard from the unsophisticated children of the wilderness were highly acceptable to Jenner, more especially when contrasted with the ingratitude of too many of his own countrymen, need scarcely be added; but let his own words express his feelings:

SIR,

Your kindness in delivering to the Five Nations of Indians my Treatise on vaccination, and in transmitting to me their reply, demands my warmest thanks.

I beg you to make known to the Five Nations the sincere gratification which I feel at finding that the practice of vaccination has been so universally received among their tribes, and proved so beneficial to them; at the same time, be pleased to assure them of the great thankfulness with which I received the belt and string of Wampum, with which they condescended to honour me, and of the high estimation in which I shall for ever hold it. May the active benevolence which their chiefs have displayed in preserving the lives of their people be crowned with the success it deserves; and may that destructive pestilence, the small-pox, be no more known among them.

You also, Sir, are entitled to the most grateful acknowledgments, not only from me, but from every friend of humanity, for the philanthropic manner in which you originally introduced the vaccine among these tribes of Indians.

I have the honour to remain, &c. &c.

E. JENNER.

Lieutenant-Colonel Gore, &c. &c.

Among the events of this period must be enumerated a scheme of his friend, Dr. Valentine, of Nancy, for honouring and recompensing Jenner. After an eloquent and animated exordium, and detailing in a brief but striking manner the history of vaccination, he alludes to the rewards conferred by Parliament, the subscriptions at Bengal and Madras, and contrasts them with the splendid and princely gifts which England had bestowed on some of her sons, whose claims were infinitely inferior to those of Jenner. He then proposes that all the societies which had been formed in the French empire for cultivating the healing art should, with the consent, and under the protection of government, open a subscription in favour of " EDWARD JENNER." The central committee of vaccination and the medical societies of the metropolis were exclusively to determine the nature of the recompense to be offered. Every learned society, every individual who belonged to the medical profession, or to any of the committees of vaccination, was to be invited to concur in this project.

After the subscriptions were closed, it was intended that deputies should be appointed who were to proceed to England to present their respectful homage to Dr. Jenner. Finally, it was proposed to erect a statue to his honour, and to place his bust by the side of Hippocrates and that of Napoleon.

This project of the kind-hearted author was forwarded to the Central Committee of Vaccination and the secretary of the Medical Society of Paris; and it was published in their Journals: but I believe it led to no result. The failure of such a scheme might have been anticipated; for had it been realized, it would have manifested a degree of enthusiasm in behalf of a foreigner that few nations have exhibited.

Though pecuniary rewards, and statues, and deputations, did not greet Dr. Jenner on this occasion, he shortly afterwards received one of the most enviable distinctions that a man of science could obtain. The National Institute of France, it is well known, have always been jealous of their reputation; and have never bestowed their honours on any one, but especially on foreigners, whose claims were not of the most unquestionable nature. On the 20th June (1808) he was elected a corresponding member of that celebrated body. I have seen a letter from a gentleman who was present at the sitting when Jenner s name was proposed; and it was received with unanimous and unequivocal demonstrations of respect.

The subject of vaccination having been treated from the beginning as subsidiary to the illustration of the character of its author, many questions that arose during its progress have been entirely omitted; because, though they deeply engaged the attention of the public when they occurred, they possess but little interest now, and do not involve in any degree points that I feel myself called upon to discuss. There are others, however, that more immediately touched Dr. Jenner's feelings, and such I have been induced to notice.

Of this description were some supposed failures of vaccination that were reported to have taken place at Ringwood in Hampshire. The affair was brought before the Royal Jennerian Society by the late Right Honourable George Rose. An excessive alarm had been created by false and exaggerated statements in newspapers, and various periodical publications. The authors of these statements were the same individuals who had distinguished themselves by a blind and inveterate opposition to the strongest evidence. In order to satisfy the public, a deputation, consisting of Mr. Ring, Mr. Blair, and Dr. Knowles, repaired to Ringwood, where they were met by Dr. Fowler of Salisbury, Mr. Rose, several magistrates and clergymen, Mr. Westcote, and Mr. Macilwain, surgeons, and the other principal inhabitants of the town and its neighbourhood. This public meeting, which was held in the Town Hall, was continued for two whole days; the medical gentlemen carrying on their investigation before this assembly in the most open manner. The result was a complete and triumphant refutation of the false and calumnious assertions that had been industriously propagated by the anti-vaccinists. This is not the place to enter minutely into particulars ; it will be sufficient to remark, that small-pox broke out at Ringwood about the middle of September : it spread rapidly and with great mortality. Vaccine inoculation did not commence until the 23rd of October, after all had been previously exposed to the contagion of small-pox. Notwithstanding these unfavourable circumstances, more than two hundred persons were successfully vaccinated, though much exposed to small-pox in different ways ; the supposed failures having occurred only in those cases where either the cow-pox infection had not taken place at all, or where the constitution had previously been impregnated with the small-pox contagion. It does not appear, that any one person either in Ringwood or its neighbourhood had caught the small-pox after going through regular and complete vaccination. Both Dr. Fowler and Mr. Rose sanctioned the accuracy of the Report, and, I believe, it satisfied every one but those who did not love the truth.

Dr. Jenner was strongly urged to form one of this deputation. This was a very unreasonable and unwise proposal; and judging from the temper of his traducers, I have no doubt that they would have made it the occasion of personal insult to him. It was actually stated in one of their publications (the Medical Observer), "that the deputies carried pistols to defend themselves against the astonished populace at Ringwood!" This

single sentence may tell the reader the nature of
that rancorous and unmanly opposition which at this
period raged against Dr. Jenner and his discovery.

The force of prejudice and error, and the evil
consequences which have resulted from them,
form the most melancholy chapters in the history of man; at one time struggling to maintain
false and inaccurate dogmas, at another resisting
the plainest and most convincing demonstrations,
we are compelled to believe that there is a principle in our nature which has too strong an affinity
for what is untrue, to permit the understanding
either to discern or acknowledge an opposite principle, till both the moral and intellectual vision
have been purified and strengthened. The per-
secutors of Galileo would, I believe, have been
eclipsed in their monstrous and outrageous hostility
to the splendid discoveries of that illustrious man, by
some of the opponents of vaccination, had the spirit
of the age or their own power enabled them to
carry their wishes into execution. It is very true
that the persons who manifested such dispositions
were little distinguished by their rank, or station,
or abilities. They were, nevertheless, men of education, and many of them belonged to the medical
profession ; and I record it as a striking proof of
the weakness of the human understanding, that in
the nineteenth century, and in the metropolis of
the British Empire, two of the most beneficial inventions should in protracted and repeated public

discussions have been consigned to contempt and obloquy, and their authors held up to the world as hypocrites and impostors. On Monday the 28th of March, 1808, the following question was discussed at the British Forum :—" Which has proved a more striking instance of the public credulity,— the gas lights of Mr. Winsor, or the cow-pox inoculation ?" The result of the discussions was as usual announced; and both vaccination and gas lights were handed over to scorn and ignominy!

This same British Forum seems to have been a place somewhat akin to that in which the Jacobins of the day put forth their pestiferous doctrines; but I have often heard Dr. Jenner aver that many individuals of our profession, and some of them, too, men holding important public stations, were concerned in diffusing such wretched and pernicious trash. I call it pernicious, merely from its influence in keeping up a resistance among the lower orders to a life-preserving practice, and in the same ratio promoting the diffusion of a most fatal disease. The walls of London were placarded with such falsehoods; and doubtless many a victim perished at the shrine of this Moloch. The same party which promoted these discussions, tried to carry their point in another way. They actually published a sort of newspaper, entitled, " The Cowpox Chronicle; or Medical Reporter." This was printed on stamped paper, and circulated through the Post Office. The wit of this publication was

very much on a level with its other qualities. I
have understood that the indulgence of their
humour became at last rather too expensive for
the proprietors. They put forth their lucubrations in the shape of advertisements, in which they
parodied all the ordinary topics that fill the
columns of a newspaper; but they were not
aware that the duty for the advertisements would
fall upon them. This weight, however, was not
required to sink the publication. Its atrocious
falsehoods, its coarse and disgusting ribaldry, and
its impious scurrility, must soon have caused its
extinction.

As a contrast to these humiliating proceedings
we may turn to the intelligence which Dr. Jenner
received from the north of Italy, from his excellent friend Sacco. He sent a long letter from
Trieste, dated January the 5th, 1808, giving an
account of his vaccinations perfectly unexampled.
"During eight years," he observes, "I reckon
more than 600,000 vaccinated by my own hand,
and more than 700,000 by my deputies in the different departments of the kingdom. I assure you,
out of a population of six millions, to have vaccinated one million three hundred thousand is
something to boast of; and I flatter myself that in
Italy I have been the means of promoting vaccination in a degree, which no other kingdom of the
same population has equalled."

The inhabitants of Italy were not insensible to

his indefatigable and disinterested exertions; and they commemorated them by ordering medals to be struck in his honour and in that of vaccination. One was executed at Brescia, another at Bologna. * These two medals, with those which had been issued on former occasions, make up eight, commemorative of the benefits conferred on mankind by vaccination; but it is a singular and somewhat distressing circumstance, that the " vera effigies " of the author of vaccination does not appear on any one of them.

Doctor Jenner's successful mediation with Napoleon for the release of persons detained in captivity is known to the reader. It is pleasing to be able to add, that the claims of science and philanthropy found favour in the breast of another crowned head, at the time when all national intercourse was suspended by the horrors of war. By a strange combination of events, a young man of the name of Powell, the son of W. D. Powell, Esq. Chief Justice of His Majesty's Court of King's Bench in Upper Canada, and nearly related to Mr. Murray, who was Secretary to the Royal Jennerian Society, was captured on board Miranda's squadron by the forces of the King of Spain; he was tried, and sentenced to ten years' confinement and hard labour in the Castle of Omoa, a sea port of Mexico. To obtain the remission of his sentence, Dr.

* See Appendix, No. III.

Jenner directly memorialised the King of Spain; and it is to the honour of all parties to be able to say, that the appeal was not made in vain. After lamenting that it could not be presented through the medium of an ambassador, he stated, that he was encouraged to hope from the magnanimity that had recently been shown by his Majesty in the glorious expedition to disseminate through every quarter of the world, alike to friends and enemies, the discovery which he had the happiness of introducing, that his petition in behalf of an unfortunate object would be received with clemency.

The circumstances of the case were then briefly detailed. The youth, only twenty years of age, was engaged in extensive mercantile concerns, and was led to the ill-fated island of St. Domingo. On his arrival he was warmly patronised by Dessalines, who encouraged him to settle, and promised him protection and support; but scarcely was he established there, before the conduct of that ferocious chief gave him cause to entertain serious apprehensions both for his property and his life. At this moment Miranda arrived with his expedition; and the imprudent youth was induced to embark in his enterprise, more with the hope of escaping the impending destruction with which he was threatened by Dessalines, than from any other motive.

Young Powell seems to have had a due and becoming feeling of the value of the intercession. He was restored to his friends at York in Upper Canada; and on the 19th day of February 1808, he, in a very grateful manner, acknowledged his obligation to Dr. Jenner.

On several other occasions he was enabled to procure the release of gentlemen who were detained on the Continent. One was the son of the celebrated Sir John Sinclair, who had gone to prosecute his studies in Germany. He was at that time at Vienna; and his friends were very anxious that he should have a passport to return through France or Holland. This object was obtained by a direct application to the Emperor of Austria, which was transmitted through Baron Jacobi. Two other gentlemen, Mr. (now Lieutenant-colonel) Gold and a Mr. Garland, were about the same time indebted for their liberation to the same intercession with the ruler of France.

At no period was the influence of Jenner's character more powerful with foreign nations. His name carried a charm with it sufficiently potent to disarm the hostility of belligerent states, and actually to turn aside the distresses and severities of war. So strong was the general feeling on this subject, that persons left our shores, not with a passport countersigned by a minister of state, but with a simple certificate bearing the name of EDWARD JENNER, testifying that the parties were known to him, and were voyaging to distant lands, in pursuit of health or science, or other affairs totally apart from the concerns of war, and were deemed by him deserving of protection and freedom from the restraints imposed upon other captives. Probably there was not a civilised nation in the world that would not have paid respect to such a document. As a curiosity I subjoin a certificate of this kind :

I hereby certify, that Mr. A. the young gentleman who is the bearer of this, and who is about to sail from the port of Bristol on board the Adventure, Captain Vesey, for the island of Madeira, has no other object in view than the recovery of his health.

EDWARD JENNER,

Member of the N. I. of France, &c. &c.

Berkeley, Gloucestershire, July 1, 1810.

On sending this certificate he adds, " I beg you to put this letter into your son's possession; and should the ship in which he sails, through the chance of war, be captured by a French commander, I trust he will, at my solicitation, shew Mr. A. every indulgence in his power; and that if he will cause it to be made known to the French government, it will obtain for him a speedy release. I feel the more confident of the Emperor's kindness in this case, as his Majesty has hitherto been pleased to lend a favourable ear to my petitions in behalf of British captives.

His applications to the Emperor had latterly

been transmitted through the hands of his physician, the celebrated Baron Corvisart, who was exceedingly punctual in making all Jenner's requests
known to His Majesty.

" J'ai remis, ces jours derniers," (he observes, in
a letter dated Paris, Dec. 5, 1809), "à S. M. la
copie de vôtre derniere lettre en date du 4 8bre
1809. L'Empereur m'a permis de vous repondre,
Monsieur, qu'il ferait mettre en liberté les deux
gentilhommes (MM. Garland et Gold), auxquels
vous vous interessez. Je suis bien flatte de pou-
voir vous annoncer cette heureuse nouvelle."

The Baron then asks Jenner to render a good
office to a young friend of his, a prisoner, who had
been sent back to France on his parole. His best
efforts were, I believe, exerted in behalf of the
young man; but, unhappily, Jenner's influence
with the British government was not equal to that
which he enjoyed with the court of France.

The adjustment of the second parliamentary
grant, the increasing importance attached to vaccination, and the decay of that institution which
had been formed for its support, called for the
adoption of other measures, better calculated to
give stability to the practice. It was, therefore,
resolved that the influence of the government
should be exerted in founding an establishment
for the propagation of vaccination throughout the
British dominions. The late Right Honourable
George Rose took the lead in this transaction.

Dr. Jenner was requested by him to draw up a
plan, and to give an estimate of the expense.
Having submitted this plan to Mr. Rose, he went
to town himself, at this gentleman's express desire, in order to assist in the organization of the
establishment. He remained five months in London, anxiously endeavouring to bring matters to a
favourable conclusion. He had frequent interviews with Mr. Rose and Sir Lucas Pepys, then
President of the College of Physicians. I will not
detail all the difficulties and mortifications that he
experienced in the negotiation, because I mean to
subjoin a paper, drawn up by Jenner, recording
his part in these transactions, and explaining the
conduct he was compelled to adopt.

During the latter part of his stay in London, his
mind was much agitated by the situation of his
family at Berkeley. His eldest son was suffering
under typhus, and he found him, on his return
home, quite a wreck, and with little hopes of recovery. His second son was seized with the
same disease, and his situation became extremely
perilous.

Sir Lucas Pepys having received the warrant
for instituting the National Vaccine Establishment, wished Dr. Jenner to be in London during
the first sitting of the board. The condition of
his family at the Chantry* kept Dr. Jenner a pri-

---

* Dr. Jenner's cottage at Berkeley.

soner there. He apologised to Sir Lucas for his absence, and begged his friend Mr. Moore to do the same, adding these remarkable words :—" In this unfortunate situation I should be unworthy of the name of father were I to stir from my children. Indeed, nothing would make me, not even a royal mandate, unless accompanied by a troop of horse." Had circumstances permitted his being in London, there is reason to fear that his feelings would have been wounded. It was natural that he should be jealous of the character of vaccination; that he should wish to see it directed by the most perfect knowledge and the purest integrity; that the bad conduct and bad faith which had been formerly displayed in other institutions should be excluded from this; that HE, the discoverer, the disinterested promulgator of the practice, should have influence enough to secure all these objects, by holding that station of dignity and responsibility in the new Institution, which certainly was his due. He was less concerned about the general arrangements of the establishment, than for the practice of vaccination itself. For *this* he felt himself responsible; and in *this* matter he very properly wished to direct.

Gentlemen who occupied prominent stations in the metropolis, could not so readily admit the claims of a provincial physician, who held no place in either of the great corporations which preside over medicine and surgery in this country. This circumstance, trifling as it might well appear to unprofessional or unprejudiced men, prevented him from being a member of that very board, which was constituted for the express purpose of promoting the practice to which he gave existence. This strange anomaly he would have overlooked, had the office assigned to himself been as efficient as its name implied. The board appointed him DIRECTOR, but they soon contrived to let him feel that he was a *Director directed.*

" It was stipulated," he observes, in a letter written on Jan. 16th, 1809, to Mr. Moore, " between Mr. Rose, Sir Lucas, and myself, that no person should take any part in the vaccinating department, who was not either nominated by me, or submitted to my approbation, before he was appointed to a station. On my reminding Sir Lucas of this, he replied, ' You, Sir, are to be whole and sole director We (meaning the board), are to be considered as nothing : what do *we* know of vaccination ?' This compact was soon forgotten ; for out of eight persons nominated by Jenner, *six* were rejected by the board. After much deliberation, he made up his mind to resign his office of Director. Some of his friends thought this step was uncalled for ; that he might have submitted to the grievances he felt, and the humiliation to which he was exposed. It is not my part to decide upon this question ; it belongs to me rather to state plainly the facts as they arose, and finally to record Dr. Jenner's own sentiments on the occasion, as contained in the following memorial, which he put into my hands a short time before his death.

"It is most painful to Dr. Jenner's feelings, when speaking of an institution, the welfare of which he cannot but have so much at heart, and the direction of which has been placed in the hands of gentlemen of such high respectability, to be obliged to express his disapprobation of the arrangements they have made, and to decline co-operating with them. But, powerful as these feelings are, he cannot suffer them to impede the faithful discharge of what he conceives to be his duty to himself, his friends, and the public.

" In the course of last year, Dr. Jenner twice went to town, in compliance with requests from Mr. Rose, for the express purpose of assisting in the formation of an establishment, the object of which was to be the propagation of the vaccine practice in the metropolis in particular, and the British realms in general. He accordingly drew up a plan for its organization, and an estimate of its probable expenses, which he laid before Mr. Rose. In the course of a few weeks he had an interview with Mr. Rose and Sir Lucas Pepys (President of the College of Physicians) together, when he found that his plan had been altered, by the introduction of a board to superintend the affairs of the establishment, by a reduction of the num-

ber of vaccinating stations, by the grant of an
annual salary to the members of the board, and by
the consequent diminution of the salaries which
were to have been granted to the officers of the establishment.   As these alterations were considered
necessary by Mr. Rose and Sir Lucas, Dr. Jenner did
not object to them, since he did not wish to interfere with the general arrangements of the establishment, but wished its affairs to be totally independent of him, except in what related to the
practice of vaccination.   But he stated at the same
time, how important it was that gentlemen should
be appointed to the inoculating stations who were
thoroughly acquainted with the practice; and that,
on this account, he hoped their nomination would be
left to him, or at least that none would be appointed who were not approved of by him.   This he
conceived to be indispensable, since the public
and the world at large would of course consider
him responsible for the manner in which the vaccine practice was conducted in an institution
with which he was connected.   Mr. Rose and Sir
Lucas Pepys acquiesced in the propriety of Dr.
Jenner's request; and that he might not only
guide the practice of vaccination, but have a share
in the management of all the concerns of the
establishment, a clause was introduced for the
particular purpose of admitting him a member of
the board.

"After remaining in London for five months, with

no other object in view than the completion of
the establishment, Dr. Jenner was obliged, by
urgent circumstances, to return to Gloucestershire.
Previously, however, to his leaving town, he
waited on Sir Lucas Pepys, and repeated what he
had formerly said of the high responsibility of his
office as Director, and of the great importance of
selecting gentlemen, for whose knowledge of vaccination he could be answerable, to conduct the
practice of it. In this Sir Lucas most readily con-
curred. Dr. Jenner, therefore, at once nominated
Mr. Moore as his assistant director, and recommended Mr. Ring as Principal Vaccinator and
Inspector of Stations, urging this recommendation
with a particular stress, since he conceived that
appointment to be of far greater consequence than
any other, on account of the extensive duties
which would be attached to it, which very few,
who would accept of the situation, would be qualified to perform, but for which Mr. Ring was particularly qualified, from his long and wide practice
of vaccination, and his intimate acquaintance with
its minutiæ. Dr. Jenner added, that when the
establishment assumed its functions, he would
send in the names of those gentlemen whom he
wished to see elected to the subordinate stations.

"After a few weeks, Sir Lucas received the warrant for instituting the establishment. The board
was formed, consisting of the four censors of the
College of Physicians, with the master and two

senior wardens of the College of Surgeons. Dr.
Jenner was not admitted a member of it, notwithstanding the clause inserted expressly for the purpose of introducing him. The board assembled to
appoint the principal officers, when Mr. Ring,
whom Dr. Jenner had so strongly recommended,
was set aside, and a gentleman appointed in his
place who was taken from an institution which
had been personally hostile to Dr. Jenner on all
occasions.

" The next meeting of the board was to appoint subordinate officers. Previous to this meeting, Dr. Jenner sent in to the board, through
Sir Lucas, the names of seven gentlemen, whom
he knew to be most eminently qualified for
conducting the vaccine practice, and whom he
therefore was anxious to see appointed. The list
of these gentlemen was presented to the board accordingly, before they commenced the nomination
of vaccinators : one was set aside, as the number
of stations was reduced by the board to six ; and
out of the six gentlemen whose names remained,
four were rejected by the board. By the whole
of these circumstances, Dr. Jenner felt himself
under the necessity of withdrawing from the establishment. He could take upon himself no responsibility where he had no power, not even a
vote. He did not wish to control the establishment; nothing was farther from his thoughts.
But he expected that the practical part of its
concerns would have been under his direction, as

the title of his office implied; and he expected
that those gentlemen, whom, from a consciousness
of their pre-eminent ability, he had so strongly recommended to conduct this practical part, would
have been appointed. But as his recommenda-
tions have been disregarded,—as arrangements
and appointments have been made which are contrary to his judgment, and as he is informed by
the board that it was intended for them to use
their own discretion, and that they alone are responsible for the conduct of the establishment,
Dr. Jenner declined accepting the station of Director to which they had nominated him, since he
found that he was to have nothing to do in the
establishment, and that his office was only a
name."

In coming to the decision mentioned in the preceding document, it was satisfactory to him that
he was enabled to keep up his communication
with the Institution through the medium of Mr.
Moore, who succeeded him as Director. He was
consulted on the first paper of instructions which
was circulated by the board for managing the
practice of vaccination; and on every occasion
wherein his assistance could be of the slightest
service it was most freely and cordially given.
Mr. Moore distinguished himself among those
friends who endeavoured to induce him to accept
the office that had been allotted to him. He
thought that Jenner took a stronger view of the

difficulties of the case than was necessary; and, with all the ardour of sincere regard, used every means to induce him to alter his purpose. A very interesting series of letters, which Mr. Moore has most kindly placed at my disposal, and some of which, in justice to Dr. Jenner, I have deemed it necessary to print in a subsequent part of this volume, will show how highly he estimated the importance of the new establishment, and how earnestly and magnanimously he laboured for its success. One of them is appropriately introduced here.

From Dr. Jenner to James Moore, Esq.

My dear Friend,

At the time I informed you of my intention to come to town, believe me I was quite in earnest. But while I was getting things in order came a piece of information from a Right Hon. Gentleman which determined me to remain in my retirement. It was as follows. *That the Institution was formed for the purpose of a full and satisfactory investigation of the benefits or dangers of the vaccine practice, and that this was the reason why Dr. J. could not be admitted as one of the conductors of it, as the public would not have had the same confidence in their proceedings as if the board were left to their own judgment in doubtful cases.* This is the sum and substance of the communication;—" What do we know of vaccination? We know nothing of vaccination."

And yet, my friend, these very *we* are to be the sole

arbitrators in doubtful cases! Alas, poor Vaccina, how
art thou degraded!

You intimated something of this sort to me some time
since, and now I get it from the fountain head. An institution founded on the principle of inquiry seven or eight
years ago, would have been worthy of the British nation;
but now, after the whole world bears testimony to the
safety and efficacy of the vaccine practice, I do think it a
most extraordinary proceeding. It is one that must necessarily degrade me, and cannot exalt the framers of it
in the eyes of common sense. I shall now stick closely to
my own Institution, which I have the pride and vanity to
think is paramount to all others, as its extent and benefits
are boundless. Of this, I am the real and not the nominal
director. I have conducted the whole concern for no
inconsiderable number of years, single handed, and have
spread vaccination round the globe. This convinces me
that simplicity in this, as in all effective machinery, is
best.

I agree with you that my not being a member of the
British Vaccine Establishment will astonish the world;
and no one in it can be astonished more than myself. An
establishment liberally supported by the British Government,—its arrangements harmonious and complete,—every
member intimately acquainted not only with the ordinary
laws and agencies of the vaccine fluid on the human constitution, but with its extraordinary or anomalous agencies,—all fully satisfied from the general report of the
civilized part of the world and their own experience of the
safety and efficacy of the vaccine practice,—all cordially
uniting in directing that practice to one grand point, the
extermination of the small-pox in the British Empire:—a
society so formed, was a consummation devoutly to be
wished. But instead of this, taking away yourself and a

few others, an assembly, which from well-known facts must appear discordant in the eyes of the public, is packed together. However, incongruous as it is, it would have been still more so, had I mingled with it; and what is above all other considerations, and which would have proved a source of perpetual irritation, I must have gone in with a sting upon my conscience.

Though resolved on not incorporating myself with the Society, be assured I shall be ever ready to afford it any assistance in my power.

Believe me, my dear Friend,

most truly yours,

EDWARD JENNER.

Berkeley,
April 4th, 1809.

The after-thought mentioned in the preceding letter, by whomsoever suggested, was quite unexpected by Jenner, and not less at variance with proper respect to himself than to the ascertained character of vaccination. The Imperial Parliament had twice, after the most mature investigation, marked its approbation of the discovery. The testimony of all the enlightened medical men throughout the world vindicated the discrimination and accuracy of the discoverer; but in the constitution of the new board, the overwhelming weight of this evidence was not deemed sufficient; and Jenner and vaccination were again to be put upon their trial. A tribunal was established which was so scrupulous in the discharge of these its inquisitorial duties, that it could not safely admit into its counsels that noble and most generous man, to whose wisdom and disinterestedness they owed their very being. I have no satisfaction in dwelling on this topic; but, in accordance with the principle which I trust has guided me throughout this work, I may not abstain from recording any facts necessary for explaining the conduct of Jenner. I am fully persuaded that he considered the Parliamentary investigation, and the Parliamentary grants too, of inferior moment to the Establishment, the origin of which I have been tracing. Such was the magnanimity of his character, that personal honours and emoluments were as nothing in his eyes when compared with the great purposes that vaccination was calculated to secure. The poor attempts, therefore, to keep down such an one ; to deprive him of the distinction which he had so well earned; to interpose the rules and forms of collegiate discipline between him and his reward, did not well consist either with the claims of science or of substantial justice. He was truly an humble man, and it afforded him no gratification to feed on empty praise. He was contented to spend his days in retirement, or even in obscurity ; but he had a supreme love of justice, and he could not condescend, for any object, to depreciate its value.

A person with such feelings, and who stood in the remarkable station which he occupied, could not maintain his self-respect, nor secure the appro-

bation of others, except by acting as he did. I
know that his trials on this occasion were searching, and hard to be endured. To see himself compelled to withdraw from a national establishment,
endowed for the purpose of propagating his discovery, was a bitter mortification to him; but his
deliberations were mature, and his decision was
firm; and I have no doubt that posterity will say
that he was right. This certainly was the conclusion of some of his warmest adherents at the time;
and in accordance with such sentiments, the late
amiable Sir Thomas Bernard thus expressed himself upon the occasion.

Wimpole-Street, 6th March, 1809.

My dear Sir,

I did not expect all that has happened; but from some
circumstances which came to my knowledge in November,
I guessed that the new Board was to be made an instrument of patronage; I therefore did not augur well of the
result. I am glad you have resigned, and have confidence, that when the Board is noticed in Parliament, the
treatment you have received will be properly censured.
It will be material to consider to whom a detail of the
circumstances should be confided. I think it will end
more for your honour, than if they had complied with
your recommendation, and you had continued director.
I wish to know when you will be in town.

With all my feelings, however, of what has recently
passed, I continue so much and so entirely gratified with
the honourable and public tribute which Parliament has
voted you, that I treat this last event, and indeed all the

other matters, as trivial, and undeserving either of your friends' attention or yours. The reflection frequently recurs to my mind, that in the great point, the national acknowledgment, there has been entire and unqualified success; and therefore, that in other matters, we may very well admit of some things not being exactly as we wish. Such are my sentiments; but not venturing to trust myself, and knowing the value you justly put on Lady Crewe's opinion, I would not answer your letter till 1 had seen her ladyship. I found her at home yesterday, and as desirous as I am that you should make your mind easy about lesser matters, and not expect the world to be composed of other materials than those experience has found in it. The number of those who honour and respect you is very great, and the adoption of your discovery *throughout the world* has been rapid and successful beyond example. Let us then not be disturbed by two or three envious calumniators, or by a few sinister events. The promulgation of every discovery by which mankind has been benefited has always been attended with similar circumstances : it is a general condition, and must be submitted to.

I shall show this to Lady Crewe before it is sent off. Let me know when we are to expect you. You will find the Alfred flourishing beyond any expectation, and in great request.

Adieu, my dear Sir,

and believe me always

most faithfully yours,

T. Bernard.

The conduct of the vaccine board was different from that of a body constituted for a similar pur-

pose in a neighbouring nation. Napoleon, in the
end of 1809, issued a decree, that one hundred
thousand francs should be at the disposal of the
Minister of the Interior, for encouraging and propagating vaccination throughout the empire.   Dr.
Valentin, who communicates this information to
Dr. Jenner, adds, " Dr. Thouret, the Dean of the
Faculty and the chief member of the Central
Committee, wrote to me, that they will not keep
for themselves any fees, as certain members in
London have done, but will employ the whole sum
allowed by the Emperor, to give rewards and
encouragements in our departments."   In another
part of the same letter he says :   " The report
made by the Central Committee to the minister,
concerning the state of vaccination in France,
during the years 1807-8, has been published
with the above decree annexed to it, and the
names of many gentlemen, physicians, and some
ladies, to whom the minister has granted gold,
silver, and brass medals for their zeal and exertions in propagating your immortal and beneficent
discovery."

When the preceding detail was written, I did
not expect that the subject of vaccination would
again come under the cognizance of a parliamentary committee.   The expediency of continuing
the vaccine board was, however, questioned in the
first session of the reformed parliament, and a
select committee was appointed to collect evidence,

and to report their opinion thereupon. The report itself, as well as the evidence on which it is founded, has necessarily attracted my attention, and I am thankful that I have an opportunity of making a few remarks on the occasion.

Before I proceed to execute this duty, I must be permitted to express my respect for the honourable members who constituted the committee. The evidence which was delivered before them doubtless afforded a sanction for the statements contained in their report. The friends of Dr. Jenner, however, may be permitted to regret that their inquiry, searching as it was in some respects, had not embraced all the points which especially affected his reputation, and the character of his discovery.

With all becoming deference, therefore, I am constrained to notice those parts which either insufficiently or inaccurately represent the occurrences which took place. No one can read what I have already said respecting the formation of the vaccine board, without perceiving that the original object of establishing that institution, as explained in Dr. Jenner's memorial, was different from that which was ultimately avowed, and which as I have already shown, was one of the main causes of his secession from the establishment. The propagation of vaccination throughout the empire, under the countenance and support of government, was the primary object of the institution. The idea of making the board a sort of court of revision, did not occur till some time afterwards, and the only effect of it was to exclude Dr. Jenner from his proper situation at the board.

The committee might not have been aware of the facts which I have stated on this point; but I hope I may be permitted to observe, that the question, whether or not vaccination was an infallible preventive of small-pox, was not involved in uncertainty when the vaccine board was established. This, therefore, could not have been a reason for constituting that board. The fact is, that the question had been nearly as completely settled at that time as at this moment.

Vaccination, and the merits of the discovery, had already been twice before parliament, and the second investigation had so fully established its value, that twenty thousand pounds, in addition to the former grant, were voted to the author. In these discussions we hear nothing of vaccination as an infallible preventive of small-pox. I am not sure that the expression was ever used by Dr. Jenner himself. If he did use it he certainly very soon accompanied it with the necessary qualification, as the quotations already printed in this volume amply testify. He may perhaps at the very outset have stated his opinion somewhat too decidedly; but no one qualified to judge of the evidence which I have already presented, can doubt that he, from the very beginning, was pos-

sessed of the gauge by which to measure the virtues of vaccination. "*Duly and efficiently performed*," he observes, "*it will protect the constitution from subsequent attacks of small-pox, as
much as that disease itself will. I never expected
that it would do more, and it will not, I believe,
do less.*" Such being the case, I am surprised
that the committee towards the conclusion of their
report should have reiterated the statement that
Dr. Jenner, at the time of the formation of the
vaccine board, announced vaccination as an "*infallible preventive of small-pox.*" "The fate of
a new practice," they continue, "was thus made
to hang on the occurrence of a single case of small-
pox after vaccination.*" This I am sorry to say is
a great misapprehension; the opposers of vaccination endeavoured to place the fate of vaccination on such an issue; but if his principles be
duly considered, he never at any time sanctioned
such an idea; and long before the practice of
vaccination became general, he anticipated failures,
and explained the circumstances under which they
were most likely to occur. I cannot help regret-
ting that evidence of this kind, which had been
before the public in various shapes for many years,
was not brought under the consideration of the
committee; if it had, ample justice would doubtless have been rendered to the memory of Jenner.

* See Report, p. 11.

The misapprehension on this point led to some mistakes in other particulars ; it gave support to the notion that Dr. Jenner's doctrines were very inaccurate, and that the whole subject required further investigation.

It would be tedious, and it is manifestly unnecessary, to repeat the evidence contained in these volumes, which proves the surprising accuracy of his first investigation, his clear and unfaltering decisions both with regard to the nature of the complaint and the method of conducting the practice. Every year's experience has only tended to strengthen his opinions, and to vindicate his character as a medical philosopher.

While some of the preceding events were in progress, my acquaintance with Jenner commenced. He was living at Fladong's Hotel, Oxford Street, in the summer of 1808, making arrangements for the national vaccine establishment. I was introduced to him at that place by Dr. Maton. I cannot refer to this and many other favours conferred upon me by this distinguished and most estimable physician without dwelling for a moment on the consequences of that introduction. To me it proved one of those leading and influential events which colour all the subsequent ways of a man's life. I was about to commence practice : all the world was before me. In seeking the acquaintance of Jenner I was impelled mainly by a desire to do homage to a man whose public and private character had already secured my warmest admiration. I little thought that it would so speedily lead to an intimacy, and ultimately to a friendship which terminated only at his death, and placed me in a relationship to his memory that no one could have anticipated. The greatness of his fame, his exalted talents, and the honours heaped upon him by all the most distinguished public bodies of the civilised world, while they made me desirous of offering my tribute of respect to him, forbade the expectation of more than such an acknowledgment as a youth, circumstanced as I was, might have expected. I soon, however, perceived that I had to do with an individual who did not square his manners by the cold formality of the world. He condescended as to an equal; the restraint and embarrassment that might naturally have been felt in the presence of one so eminent vanished in an instant. The simple dignity of his aspect, the kind and familiar tone of his language, and the perfect sincerity and good faith manifested in all he said and did, could not fail to win the heart of any one not insensible to such qualities. Though more than twenty years have elapsed since this interview took place, I remember it, and all its accompaniments, with the most perfect accuracy. He was dressed in a blue coat, white waistcoat, nankeen breeches, and white stockings.* All the tables in his apartment were

* We are grateful to him who told us that Milton wore

covered with letters and papers on the subject
of vaccination, and the establishment of the National Vaccine Institution. Having recently come
from Edinburgh, he talked to me of the excellent article which had lately appeared in the
Edinburgh Review, relative to the vaccine controversy.* He spoke with great good humour

large buckles; and that Washington broke in his own
horses; and in some future day the curious reader may be
thankful for such particulars descriptive of the habits of
Jenner.

* The paper to which he alluded on this occasion was
one of the most impressive that had been published since the
promulgation of the vaccine discovery. Its intrinsic value
was great, and its influence must have been felt wherever it
was duly examined. It acquired, however, an added virtue
from having been printed in a work which had obtained a
most extensive circulation and powerfully affected the public
mind. There were other circumstances favourable to the
cause which it advocated. It did not appear in a medical
work; it was not written by a medical man, and it kept
aloof from every thing like professional prejudice or party
violence. It exhibited an eloquent and masterly analysis of
the various points at issue between the different combat-
ants; it clearly and forcibly balanced their arguments, and
so convincingly and manifestly shewed on which side truth
and justice lay, that all reasonable minds were satisfied.

Several of the first sentences are worthy of a place in this
work. I do not know that on any occasion the benefits
arising from vaccination have been placed in a more striking
and impressive aspect. " Medical subjects," it is observed,
" ought in general, we think, to be left to the medical journals; but the question as to the efficacy of vaccination is of

also of the conduct of the anti-vaccinists, and
gave me some pamphlets illustrative of the con-

such incalculable importance, and of such universal interest,
as to excuse a little breach of privilege. We let our lawyers
manage actions of debt and trespass as they think proper,
without our interference; but when the case touches life or
reputation we insist upon being made parties to the consultation, and naturally endeavour, at least, to understand the
grounds of the discussion. The question now before us is
nothing less than whether a discovery has actually been
made by which the lives of 40,000 persons may annually be
saved in the British Islands alone, and double that number
protected from lengthened suffering, deformity, mutilation,
and incurable infirmity! This is not a question, therefore,
which is interesting only to the physiologist or the medical
practitioner; it concerns nearly every community in the
universe, and comes home to the condition of almost every
individual of the human race; since it is difficult to conceive
that there should be one being who would not be affected by
its decision either in his own person, or by those of his
nearest connexions. To the bulk of mankind wars and revolutions are things of infinitely less importance; and even
to those who busy themselves in the tumult of public affairs
it may be doubted whether anything can occur that will
command so powerful and permanent an interest, since there
are few to whom fame or freedom can be so intimately and
constantly precious as personal safety and domestic affection."

I have already had occasion to allude to the character of
the vaccine controversy; but my delineation was feeble when
compared with the authoritative declarations of a writer well
acquainted with literary and scientific warfare. "In the
whole course of our censorial labours we have never had occasion to contemplate a scene so disgusting and humiliating
as is presented by the greater part of this controversy; nor

troversy then carrying on. The day before I
saw him he had had an interview with the Princess of Wales, and he shewed me a watch which
Her Royal Highness had presented to him on that
occasion. I did not see him again till the following year, after I had fixed myself at Gloucester.
Our second interview took place in his own house
at Berkeley. His eldest son Edward was then
lying in the last stage of pulmonary consumption.
He had repeated hæmorrhages from the lungs,
and was then evidently approaching his end. Dr.
Parry of Bath was in the house. I was introduced
into the sick-room, and there for the first time saw
Mrs. Jenner, the anxious and constant attendant
on her dying child. Jenner was particularly attached to this young man, and apparently for
qualities which, in less generous natures, would
have produced a different effect. He had been

do we believe that the virulence of political animosity, or
personal rivalry, or revenge, ever gave rise among the lowest
and most prostituted scribblers to so much coarseness, illi-
berality, violence, and absurdity as is here exhibited, by
gentlemen of sense and education, discussing a point of professional science with a view to the good of mankind." The
writer, notwithstanding this discouragement, proceeds with
his task ; and in a most able and luminous manner seizes upon
all the strong points of the argument, and, after weighing the
evidence in a most impartial manner, comes to the conclusion
that vaccination, if it do not absolutely and certainly secure
the patient from the contagion of small-pox, gives him a security at least as effectual as could be given by the old practice of inoculation.—See Edinburgh Review, vol. ix. page 63.

always delicate in health, and had, moreover, in some respects, rather a defective understanding. His father felt these deficiencies, and considered them but as stronger claims to his attachment and his regard. Some years after, he wept when he talked to me of this son, and many times referred to the singular character of his mind with the most touching and affectionate recollections.

Poor Edward lingered to the beginning of the year 1810. A letter written by Jenner to his friend Mr. Hicks contains strong proof of the intensity of his feelings and the depth of his sufferings. "I feel," said he, "greatly obliged to every one who attempts to console me in my present affliction; but you, who know so much of the human mind, are convinced how vain are these friendly efforts. I had no conception till it happened that the gash would have been so deep; but God's will be done! In the midst of my wretchedness a ray of comfort sometimes breaks in upon me and tells me my sorrowing will be profitable to me. How mysterious and unsearchable are the ways of Providence! God bless you.

Your affectionate

E. JENNER."

Berkeley, Feb. 10*th*, 1810.

"One would suppose," he observes, writing to another friend [John Ring], "that the mind would become in some measure reconciled to an event,

however melancholy in its nature, that one knows
to be inevitable, when it has made such gradual
approaches ; but I know, from sad experience, that
the edge of sensibility is not thus to be blunted."
But his domestic sorrows did not materially impede his efforts in the cause of vaccination. They
were less interrupted, however, by the attacks of
his opponents. After using the expressions just
quoted, he observes, " Whether it be from age,
long retirement, or what I cannot tell, but some
how or other, I feel myself less and less disposed
to notice the malevolence of my enemies ; and as I
wish *you* quite as well as I do myself, I should be
happy to see you follow my example, entirely on
the principle of your enjoying more repose. The
world is ungrateful, and will never requite you for
your toils. Your satires against the anti vaccinists
are keen ; but the keenest of all are those which
you engrave with the point of your lancet. Where
is the man in a private station in the metropolis
who has rescued so many from an untimely grave
as yourself ? This is a satire which must deeply
wound the hearts of our opponents if they are not
too callous to be penetrated. The paper inclosed
is another satire on their absurdities. Would not
some of the editors of the newspapers be glad to
insert it ? The Duke of Sussex was some time
since taken very ill at Gloucester. I attended His
Royal Highness there ; and afterwards, for near a
fortnight, at Berkeley. This gave me an opportunity

of bringing up Corneiro, and of going fully into the subject of his abominable pamphlet. The Duke understood the matter so well that he could refute every charge which related to Portugal, and explained many things very minutely, particularly that respecting the royal infant, who he assured me was vaccinated (I think he said by Domeyer) in spite of all remonstrances." A short time after this event His Royal Highness, in testimony of his respect for Jenner, sent him by the Countess of Berkeley a very handsome hookah. In returning thanks for this very elegant present, Jenner ob served, "Your Royal Highness's kindness in making this addition to the scanty number of gratifications afforded me in this sequestered spot, will never be forgotten; and smoking, I am sure, is a harmless one if used in moderation. A man who has a pipe at his command, independently of its salutary influence in some instances, has always a soothing companion.

"It was with great pleasure I heard that your Royal Highness was enjoying so good a state of health. Guard well, Sir, your stomach and your skin, and I am persuaded you will live in safety from the future attacks of the malady that so often annoys you. Happy shall I be, if at any time I can do any thing to convince you how much I am

your Royal Highness's

obliged and devoted

humble servant,

EDWARD JENNER."

The shock produced by the death of his son, and his incessant labours, materially affected his health. Writing on the 1st of April to Sir Thomas Bernard, he says, " Your letters are always pleasant to me. I was in your debt when you were good enough to send the last, and should have answered it long ago, but for a most afflictive event which has happened in my family—the death of my eldest son, an amiable youth, who had just reached his twenty-first year. This melancholy occurrence threw me into that state of dejection which renders me unfit to perform my ordinary duties, and I still feel nveloped as it were in clouds, so that all objects wear a new and gloomy aspect. You wish me to come to town; you will find me too torpid to perform any useful offices; and I feel confident that even the cheerful company of yourself and those friends into whose society you have so often introduced me, would at present do me no service I bend to the will of Providence, trusting in due time that I shall from this source derive that consolation which no other can afford."

His symptoms became so distressing, that active means were deemed necessary to obviate them. In pursuance of this object he went to Bath. I was to have accompanied him, but was disappointed in my wishes. The following letter, which he wrote me on his return to Berkeley, is at once illustrative of his bodily and mental state

My dear Doctor,

The Bath scheme would have turned out well if it had
not been for an unexpected *chasm* which occasioned a
general disappointment; but to none of the party so particularly as myself, as I thought to have benefited by the
union of your medical ideas with my friend Parry's, on
my case. I have been cupped, calomeled, salted, &c. &c.
and think the cascades do not roar so loud in my ears as
they did, nor my head feel so heavy; but still all is far
from right. The constant disposition to drows'ness is a
lamentable tax upon me. I should not say constant, but
frequent, for I feel lively from breakfast time till about
one o'clock, when the signal of acidity in my stomach
is the signal also for nodding. For six days my only
drink was water; but finding my pulse sink from its old
standard, 48 to 40, I venture again on two glasses of wine
after dinner. I am ordered to migrate, and not to think.
The first injunction I shall comply with, and go to town
on Monday for a few days. On my return it is my intention to throw myself upon Cheltenham in a probationary
way. Should this be too much for me, my retreat is not
far distant.

You see I have enlisted Creaser in our cause, and you
will find that he will not discredit it. The inclosed is
from Filkin, surgeon of the Gloucester regiment of Militia. We must not repine at the ill success he has met
with. He is on that barren kind of land where science
has not yet begun to vegetate. Poor Chatterton, by
accident, sprang up on this soil; but he was soon rooted
up and flung away like a weed.

Our friend Hicks goes with me to town; I wish you were of the party.

Adieu!

Most truly yours,

EDWARD JENNER.

I find no proof of the torpor of which he complains above in his correspondence at this season. On the contrary, many of his letters, though tinged by the pensive hues of his affliction, are full of energy and vigour when the subject required him to put forth his strength.

Immediately on his return to Berkeley he had another painful duty to perform in watching the progress of the fatal disease which put a period to the life of the late Earl Berkeley. During this attendance he had occasionally the aid of his friend, Dr. Parry of Bath. This intercourse led to a renewal of many of their ancient recollections. Some of Dr. Jenner's pathological views necessarily became subjects of discussion, and I am very glad to find that so excellent and accomplished a judge as Dr. Parry did justice to his acquirements in this branch of knowledge. In a letter which is now before me, Dr. P. observes, " It will give me great pleasure to hear your pathological theories, because, without flattery, I highly respect them all, and have no hesitation in saying, that it is your own fault if you are not still the first pathologist existing." He then goes on to remark in refer-

ence to the anti-vaccinists :—" For heaven's sake,
think no more of these wasps, who hum and buz
about you, and whom your indifference and silence
will freeze into utter oblivion. Let me again
entreat you not to give them one moment's consideration, *opus exegisti ære perennius.* The great
business is accomplished, and the blessing is ready
for those who choose to avail themselves of it;
and with regard to those who reject it, the evil
will be on their own heads." (December 8th
1810.)

Dr. Jenner was released from his attendance at
Berkeley Castle, by the death of the Earl, in the
beginning of August. This distressing event was
felt by him with great poignancy, and it materially
deepened his other sources of suffering. Towards
the end of September he changed his residence
with his family to Cheltenham, and he had been
but a short time there when he was stricken by
another calamity in the death of his sister, Mrs.
Black. This lady, and his elder brother, stood to
him almost in the relation of parents, and he
mourned her loss with the greatest sincerity.

I was in frequent intercourse with him at this
time, and had many opportunities of witnessing
the distress and embarrassment occasioned by
these domestic trials, and by the pressure arising
from the great responsibility which he never
ceased to feel respecting vaccination. Though,
as far as silence was concerned, he could follow

Dr. Parry's advice, yet he could not altogether escape from the annoyance occasioned by the blindness and wickedness of his traducers. The combined influence of all these causes sometimes rendered duties which he had formerly executed with ease, burdensome to him, and in this state it may readily be conceived that he would be willing to shrink from all labours that were not of a very pressing nature. He was of course disposed to avail himself of the help of his friends to relieve him from the burden, and he more than once honoured me by accepting my services in this way.

Dr. Jenner was an original member of the Medico-Chirurgical Society. The first volume of their Transactions contains two papers which he contributed. One on the distemper in dogs was read on the 21st of March, 1809. This curious affection, according to his observations, had not been known in this country much above fifty years. The paper is short, but contains a very faithful description of the disease. It does not, however, as might have been expected, record any of his experiments instituted with a view to ascertain the effects of vaccination in preventing the distemper. There is great reason to believe that this influence is considerable. A friend of Dr. Jenner's, Mr. Skelton, a sporting gentleman in Yorkshire, subsequently made some very decisive trials. " Having selected," he observes, " three couples of healthy pups of six weeks old, I inoculated three of them with the cow-pox under the left arm, a little above the elbow, which regularly matured. The other three with those inoculated were sent out to quarters. At a proper age they were all brought to the kennel. The former with other hounds were soon attacked with and died of the distemper, whilst the latter remained perfectly healthy, though surrounded by their infected companions, becoming the strongest hounds in the pack, and having certainly the best noses."

The other paper published by Dr. Jenner related chiefly to secondary small-pox. The facts which he had observed led him to infer that the susceptibility to receive variolous contagion remains through life, but under various modifications or gradations. His principal design in publishing these observations was to guard those who may think fit to inoculate with variolous matter after vaccination, from unnecessary alarms ; a pustule may sometimes be thus excited on those who had previously gone through the small-pox, febrile action of the constitution may follow, and, as has been often exemplified, a slight eruption. At the commencement of vaccination he deemed this test of security necessary ; but he felt confident that we have one of equal efficacy, and infinitely less hazardous, in the re-insertion of the vaccine lymph.

Notwithstanding all the evidence that had been

collected, a great deal of wilful ignorance still remained regarding the character of vaccination. Rumours and assertions of the most unfriendly nature were eagerly circulated, and much more industry was exerted in the propagation of small-pox than could have been believed. There can be no doubt that the hesitation and indecision of medical men contributed essentially to this evil; and it was imagined that a great moral duty, as well as a strong professional obligation, required that all well-informed and respectable medical men should take a decided line of conduct touching the question which then so divided the public mind. Under this conviction an association of the faculty was formed in the county of Gloucester, the great object of which was to promote the vaccine, and to discourage the small-pox inocula-tion. The members proved their sincerity by voluntarily and publicly renouncing the latter, and that solely and entirely from a thorough conviction of the efficacy of the former. A short address, em-bodying the argument for adopting this principle, was circulated through the district, and very cor-dially and generally received. It fell to my lot to take an active part in this transaction. The utter renunciation of small-pox was, even at that time, considered a bold measure, and some could not readily perceive that it was the immediate and necessary consequence of the approval of vaccina-tion. They could not discover that to disseminate the small-pox with a full knowledge that they had it in their power to avert this scourge, was an act neither quite consistent with high professional feeling, nor with their duty to their fellow-creatures. These principles found a most zealous and firm supporter in my late colleague, Charles Brandon Trye,* Esq. F.R.S. Senior Surgeon to the Glouces-

* When co-operating with him for that purpose, I little thought that I should have so soon to deplore the termination of his valuable life. I allude to this occurrence here, because it brought me for some days into close and constant professional attendance with Dr. Jenner. He had been sent for from Cheltenham, on the first approach of Mr. Trye's illness. Unhappily it wore from the very beginning a most unfavourable aspect; it began on the 3rd of October, and terminated early on Monday the 7th (1811). During nearly the whole of that time, Jenner's services were unremitting.

Trye was a man who had acquired a very high reputation in his profession ; he had a heartfelt respect for Jenner, and the feeling was mutual. He was descended from a very ancient family. He is well known to the profession by several works, and he acquired great celebrity for his skill and success in performing some of the most difficult operations in surgery. But I am not so desirous of pointing the reader's attention to such things, as to some other parts of his character which are less common. He was remarkably devoid of ostentation, and was deeply imbued with a profound and fervent spirit of devotion.

He never, I believe, performed an operation without retiring to meditate and pray, and to seek for guidance and assistance. He had committed many of his prayers to writing, and so discreet and secret was he in this matter, that it

ter General Infirmary. This was not the only
occasion in which he rendered efficient service to
the cause of vaccination. His decided and manly
character gave a powerful impulse to every cause
that he espoused, while his extensive professional
acquirements, and his eminent skill and dexterity as an operator, made his authority respected
throughout the district in which he lived. Were
this a fit occasion I would insert in this place his
first published letter on the subject of vaccination.
It conveyed all the caution of the philosopher with
the warm, unmixed approbation of a discovery as
remarkable in its origin as it was important in its
results. He sincerely respected Jenner; but neither private friendship nor any other motive could
ever touch his mind, so as to bias his judgment on
an important professional topic. He was the
plainest and most straightforward man I ever met
with. This occasioned an apparent abruptness of
manner: but there was truth in all he said and
in all he did.

Jenner was particularly pleased with our association: it afforded at once a test of the sincerity
of those who professed themselves friendly to vaccination; it offered the best arguments against the
practice of small-pox inoculation; and promised,
if conscientiously followed up, to extinguish

was not known he had been thus employed till his papers
were examined after his death.

that pest altogether. We fondly hoped that the maxims and motives that governed us would extend themselves, and that the whole moral and medical influence of the profession might be brought to bear in such a way on the prejudices of the public as ultimately to overthrow them. It had been well for the health of the community had our design prospered; for it is unquestionable, that one of the main causes of the continuance of small-pox among us arose from the ambiguous conduct of those medical men who thought it no sin to employ either small-pox or cow-pox, as it might suit the caprice of their patients.

About this period, Dr. Jenner was somewhat astonished by hearing of a notice of a motion in the House of Commons for a Bill to prevent the spreading of the small-pox. This bill was introduced by the late Mr. Fuller. It was prepared without any communication with Dr. Jenner. Suspecting that the provisions might not have been very well considered, he posted to town to inquire into the matter. Mr. Fuller very kindly put the Bill into his hands for amendment. It was returned with the objectionable parts removed; but to the great astonishment of Jenner, it was introduced to the House in its original form, and received, as he predicted, with aversion, and completely failed in its object.

At the time that Mr. Fuller was endeavouring

by legislative enactments to restrain the spreading
of small-pox, several private individuals were
attempting to accomplish the same object. Mr.
Bryce, of Edinburgh, put forth a plan of this kind.
It was much approved of by Dr. Jenner, but like all
others with a similar view, was rendered difficult
of execution by the liberty claimed by the people
of this country, of following the bent of their own
inclination.

CHAPTER V.

CASE OF THE HONOURABLE ROBERT GROSVENOR—JENNER'S
SENTIMENTS ON THAT SUBJECT—HIS ATTEMPTS TO LIBERATE
A FRENCH PRISONER—ELECTED FOREIGN ASSOCIATE OF THE
NATIONAL INSTITUTE — SIR JOSEPH BANKS AND THE
ORIGINAL PAPER ON VACCINATION—POEMS IN HONOUR OF
VACCINATION.

DR. JENNER had himself a considerable illness
in the spring of 1811, and he moreover had the sorrow of witnessing the advances of a painful and protracted disease in the person of his late brother-in-
law, Thomas Kingscote, esquire. He had occasion
to be in town early in March ; but, I believe, it
was only for a very few days. Affairs of a more
urgent nature required his presence again in the
capital, and he went there in the first week of June,
1811. On his arrival he encountered one of the most
unpleasant events that had befallen him in his
vaccine practice. Alleged cases of small-pox after
vaccination, had been reported before. Some of
them, doubtless, were real ; but the majority unquestionably were distorted exaggerations. It was
then, as it is now, that careful and skilful conductors
of the vaccine practice met with few disappointments,

while in the hands of others they were frequent.
Dr. Jenner had for thirteen years been carrying
on vaccine inoculation on a very extended scale.
Of the many thousands who had been subsequently
exposed to the influence of small-pox, not one was
known to have been affected by that disease. I
have already shown in the former volume that the
statements touching the absolute protection afforded
by vaccination were too unconditional ; but, certainly,
Dr. Jenner's own experience, up to the time we are
now speaking of, did afford countenance to these
statements.

On the 26th of May, 1811, the Honourable Robert Grosvenor was seized with symptoms which
denote the approach of a very violent disease. In
four days he became delirious, and an eruption
appeared on the face. At this time the existence
of small-pox was not suspected, because he had been
vaccinated by Dr. Jenner about ten years before.
In the course of the following day, however, the
eruption increased prodigiously, and some of the
worst symptoms of a malignant and confluent small-
pox showed themselves.

Master Grosvenor was attended by Sir Henry Hal-
ford and Sir Walter Farquhar. These gentlemen,
in making their report of this case, observed that
they entertained a most unfavourable opinion of the
issue of the malady, having never seen an instance
of recovery under so heavy an eruption. It seemed,
however, to use their own words, " that the latter
stages of the disease were passed through more
rapidly in this case than usual; and it may be a

question whether this extraordinary circumstance as well as the ultimate recovery of Master Grosvenor, were not influenced by previous vaccination."

Dr. Jenner happened to be in London when this distressing disease was in progress, and he visited the patient, in company with Sir Walter Farquhar, on the 8th of June. By that time the disease was on the decline, and the symptoms, which but a few days before threatened a fatal termination, had begun to disappear. The other children of the earl, who had been vaccinated in 1801, were exposed to the contagion of the small-pox, under which their brother was suffering, and were also subjected to small-pox inoculation, without effect.

This disastrous occurrence in the family of Lord Grosvenor, and several others of the same description, which took place about this time, induced the board of the National Vaccine Establishment to publish a special report. This measure, prudent and judicious in itself, was rendered necessary in consequence of the great interest attached to the subject. The reader who has attended to the previous discussions in this work, will find no difficulty in explaining events of this kind. The possibility of small-pox succeeding small-pox having been ascertained, no one need have been surprised that the same disease might occasionally succeed cow-pox. Dr. Jenner at once admitted the failure, and gave the true explanation of the occurrence.* The se-

* It will be seen in another page that Dr. Jenner had not been perfectly satisfied with the progress of vaccination in this child.

verity of the disease in this case afforded an exception to what generally happens, the symptoms being usually exceedingly mild. The power of vaccination was not the less remarkable : for at the very time that the greatest danger is observable, and when death is most likely to ensue, the disease rapidly abated. The complete resistance also of all the other children to the small-pox contagion, demonstrated that the failure was to be ascribed to that peculiarity of constitution which probably would have left the patient exposed to a second attack of small-pox, had he previously had the disease.

It was some time, of course, before these truths were duly appreciated. The immediate tendency of the event in Lord Grosvenor's family was to occasion no small alarm and consternation in the minds of those whose children had been vaccinated. Dr. Jenner was obliged to attend and render explanations to many persons in this state of doubt. He had many letters to answer on the same subject; some of them written with great earnestness and apprehension. Many persons immediately put the powers of vaccination to a test by having their children inoculated with small-pox. I cannot better explain this whole matter than by inserting some of Dr. Jenner's own sentiments written at the time of the event.

To Miss Calcraft.

Take a comprehensive view of vaccination, and then ask yourself what is this case ? You will find it a speck, a mere microscopic speck on the page which contains the history of the vaccine discovery. In the very first thing I wrote upon

the subject, and many times since, I have said the occurrence of such an event should excite no surprise; because
the cow-pox must possess preternatural powers, if it would
give uniform security to the constitution, when it is well
known the small-pox cannot; for we have more than one
thousand cases to prove the contrary, and fortunately seventeen of them in the families of the nobility. We cannot
alter the laws of nature; they are immutable. But, indeed,
I have often said it was wonderful that I should have gone
on for such a series of years vaccinating so many thousands,
many under very unfavourable circumstances, without meeting with any interruption to my success before. And now
this single solitary instance has occurred, all my past labours
are forgotten, and I am held up by many, perhaps the majority of the higher classes, as an object of derision and contempt. There is that short-sightedness among them (I
will not use a harsher term) which makes them identify a
single failure with the general failure of the vaccine system.
Before their dim eyes stand two cases in the family of Lord
Grosvenor, which they cannot see, or will not. There are two
children vaccinated ten years ago, who have been constantly
exposed to the infection of the other child, and inoculated
for the small-pox also; but all without effect. The infected
child would have died,—that is universally allowed,—but
for the previous vaccination. There was but little secondary
fever; the pustules were much sooner in going off than in
ordinary cases; and, indeed, the whole progress of the disease was different. It was modified and mitigated, and the
boy was saved. What if ten, fifty, or a hundred such events
should occur? they will be balanced an hundred times over
by those of a similar kind after small-pox. This is what I
want to impress on the public mind; but there will be great
difficulty in bringing this about because the multitude decide
without thinking. No less than three cases of this description have happened in the family of one nobleman (Lord
Rous). But I must check myself, lest I should tire you by

going too far into the subject. I should not have said so much, had it not appeared to me that even your judgment was carried down the tide of popular clamour. I beg my compliments to Mrs. St. Quintin. I dare say her children are very secure; but, if she has the weight of a feather on her mind, the safest and best test is vaccination with matter taken in its limpid state. I have stated my reasons for this over and over, in print and out of print.

June 19th.

In conjunction with the preceding extract, I subjoin one more from a letter addressed to Mr. Pruen, which has a direct bearing on the subject. It is valuable also, as affording another explicit testimony respecting the accuracy of the opinions delivered in the former volume. " In the eye of philosophy, or indeed of common sense, the failures that have happened in so great a mass of vaccinations are totally unworthy of serious attention. They should call forth the inquiry of the faculty to discover the cause if possible, but not their clamour; and as for the public, I think your decision in supposing them fit arbitrators in such a case as that which has appeared, is erroneous. They know no more of the laws of the animal economy than those of Lycurgus. I have ever considered the variolous and the vaccine radically and essentially the same. As the inoculation of the former has been known to fail in instances so numerous, it would be very extraordinary if the latter should always be exempt from failure. It would tend to invalidate my early doctrine on this point."

Another letter written to myself on this subject

puts the question in a different light, and is therefore worthy of notice.

Cockspur-street, Charing Cross,
June 11th, 1811.

MY DEAR FRIEND,

I should be obliged to you to send me, by the first coach,
some of the Reports of our association. It will probably be
my unhappy lot to be detained in this horrible place some
days longer. It has unfortunately happened, that a failure
in vaccination has appeared in the family of a nobleman
here; and, more unfortunately still, in a child vaccinated by
me. The noise and confusion this case has created is not to
be described. The vaccine lancet is sheathed; and the long
concealed variolous blade ordered to come forth. Charming!
This will soon cure the mania. The Town is a fool,—an
idiot; and will continue in this red-hot,—hissing-hot state
about this affair, till something else starts up to draw aside
its attention. I am determined to lock up my brains, and
think no more *pro bono publico;* and I advise you, my
friend, to do the same; for we are sure to get nothing but
abuse for it. It is my intention to collect all the cases I
can of small-pox, after supposed security from that disease.
In this undertaking I hope to derive much assistance from
you. The best plan will be to push out some of them as
soon as possible. This would not be necessary on account
of the present case, but it will prove the best shield to protect us from the past, and those which are to come.

Ever yours,
EDWARD JENNER.

The tone of the preceding letter shows a soreness
somewhat unusual with him. The disaster in Lord
Grosvenor's family had powerfully excited the public
mind, and roused the anti-vaccinists to increased

efforts. But Jenner was ill at ease on other subjects.
He had been required to give evidence before the
House of Lords respecting the Berkeley Peerage;
and this, added to a clamour in high places on the
subject of vaccination, would easily account for the
acuteness of his feelings. An exhibition before any
large assembly was always to him an embarrassing
and perplexing affair. His own account of the
species of mental torture that he endured when
preparing for the annual festival of the Royal Jennerian Society, as stated to myself, will render it less
a matter of wonder that he should have experienced
an unusual degree of perturbation when before the
assembled nobles of the land.

"I can compare my feelings to those of no one
but Cowper the poet, when his intellect at last gave
way to his fears about the execution of his office in
the House of Lords. It was reading Cowper's Life, I
believe, that saved my own senses, by putting me fully
in view of my danger. For many weeks before the
meeting I began to be agitated, and, as it approached,
I was actually deprived both of appetite and sleep;
and when the day came, I was obliged to deaden my
sensibility and gain courage by brandy and opium.
The meeting was at length interrupted by a dissolu-
tion of Parliament, which sent the leading people to
the country; and what was at first merely postponed
was ultimately abandoned, to my no small delight
and satisfaction."

Notwithstanding the unfavourable occurrence in
Lord Grosvenor's family, vaccination continued to
flourish, and he had the happiness to learn that the

deaths from small-pox in the bills of mortality were at this time reduced to five or six a week. This reduction was in a great degree to be ascribed to the well-directed efforts of the National Vaccine Establishment. It will scarcely be believed in future times, that small-pox inoculation still continued to be practised at some of the principal establishments for the relief of the sick poor in London. In that number, I am sorry to be obliged to mention the Finsbury Dispensary. The Board of the National Vaccine Establishment exerted their influence with the directors of that institution, and the practice was abandoned, but it was still continued in the Small-pox Hospital!!!

I find a curious document, stating the expenses incurred by Dr. Jenner up to this time, in consequence of the two Parliamentary grants, and other contingencies immediately arising from his discovery. Those connected with the first grant of ten thousand pounds, amounted to no less a sum than £977. The grant in 1807 was better managed: only £58. 18s. 10d. being required. The other sums which he expended in printing, &c. amounted nearly to £700. If to these we could add the numberless other sources of expenditure, or even one of the items, such as postage, without counting the value of his time and labour, it would be found that a very formidable amount would stand against the public grants.

Dr. Jenner was enabled to leave London early in July, 1811. The annoyances he had recently encountered, the state of Mrs. Jenner's health, and

that of her brother in Cheltenham, had an unfavourable effect on his own. I remember finding him
seriously distressed likewise by the fate of a young
French officer, in whose welfare he took a deep interest.

Captain Husson belonged to Dupont's army, and was
one of those who capitulated after the defeat of that
General at Baylen. Jenner felt for young Husson as
for a son; and exerted himself with unremitting earnestness to procure his liberation. Unfortunately,
as I have already stated, his influence with the British
Government was not so great as it was at one time with
that of France. Either by direct appeals to the Emperor, or through the medium of Corvisart and Husson,
he had succeeded in procuring the release of many
persons. " This," he observes to a friend, " though
somewhat gratifying to my feelings, will convince
you of the feebleness of my influence; and feeble as
it has ever been, it has now become more so from the
following unfortunate occurrence. M. Husson, one
of the medical gentlemen above referred to, has a
brother, Captain Husson, a prisoner of war in this
country. I petitioned for his release. It was the first
request of the kind I had made to the British Government; and it seemed to meet with a favourable
reception. This joyful intelligence I communicated
to Captain Husson: when, most unexpectedly, instead of acquiescence, a refusal arrived. This threw
him into a state of desperation, and in the midst of
it he broke his parole, and is now in a state of misery
and confinement on board a prison-ship at Chatham.
This hurts me excessively: as I cannot but look on

myself as in some measure the innocent cause. I
shall again exert myself for Captain Husson, conceiving that my first application could not have been
well understood." The second application, I believe,
was successful. At least, he concludes a letter to
myself dated the 5th of April, 1811, in these words.
"This day I received a letter from town, informing
me that my petition to the Prince in behalf of Captain Husson had been graciously received."

I trust, if M. Husson or his brother should ever
see these pages, they will believe that Dr. Jenner,
whatever may have been the event of his application,
was sincere and unremitting in his efforts; although,
while overwhelmed by domestic affliction, or the
labours connected with his peculiar station, he may
have sometimes omitted to reply to the letters of M.
Husson: but for reasons which are obvious, I wish
to make it manifest, that he did not forget the claims
which M. Husson and his countrymen had upon
him for his best services. I am a witness, that such
services were rendered; and farther, to remove any
blame that might be attached to Dr. Jenner for his
silence, I myself, at his request, wrote to M. Husson,
explaining his difficulties, stating what he had done,
and was doing in behalf of his brother; and that he
was not unmindful of the great benefits conferred
upon himself and the cause of vaccination by the
French nation.

I have an anxious desire that this matter may be
well understood by all whom it may concern. From
the altered manner of M. Husson, in speaking both
of Dr. Jenner and vaccination in the Dictionnaire des

Sciences Medicales, which I have been obliged to refer to very pointedly in the former volume, I fear that he may have been imperfectly informed on this subject, and believed Dr. Jenner alike insensible to the calls of gratitude and of duty. Indeed, in a letter which is now before me, he seems almost to reproach Jenner, as if he really had not exerted himself with the British Government so effectually as he might have done. He thought it not credible that they could refuse to the author of vaccination what the French government had granted to him. Little did he know how the prejudices of our rulers interposed on all occasions when a question arose respecting any dealings with the person who then wielded the destinies of France. And, moreover, so small was Jenner's interest with those in authority, who dispensed favours in this country, that he never succeeded in obtaining an appointment for either of his nephews, or any other of his connexions. When recounting the ill-success of his many applications on their behalf, I have heard him good-humouredly remark, that he had once got a place for an exciseman, but nothing beyond it.

In 1808 the National Institute of France had elected Jenner a corresponding member. The same distinguished body conferred a still higher honour upon him by placing him, on the 13th of May, 1811,* in the list of foreign associates. This vacancy was occasioned by the death of Dr. Maskelyne. The approbation of the Emperor and King was accorded

* This honour was conferred on the vaccination of the King of Rome.

to their choice at the Palace of Rambouillet, on the
19th of the same month.   Sir Joseph Banks, in announcing the honour, says—

MY DEAR DOCTOR,

I have great pleasure in transmitting to you the notice of
your election in the National Institute of France, in the place
vacated by the demise of Dr. Maskelyne, and in congratu-
lating you on your obtaining a place among a body of men
who have so little humbled themselves before the arbitrary
dispositions of their Sovereign as to have retained the title of
National, when that of Imperial was offered to them, and
who, I verily believe, are as little satisfied with the barbarous
mode of warfare adopted by their chief, as we Englishmen
can be.

Adieu, my dear Doctor,
Always faithfully yours,
JOSEPH BANKS.

Soho-Square,
Dec. 4th, 1811.

On looking at the diploma forwarded by Delambre
to Dr. Jenner, I observe a curious proof that the influence of the Emperor had, at least in this instance,
prevailed over the firmness of the members of the
Institute.   The engraved document was headed
" INSTITUT de FRANCE."   In the copy before me,
the abbreviation IMP. in a small hand is inserted
after the word INSTITUT.

The cordial manner in which Sir Joseph communicated this flattering mark of distinction, and his
conduct on other occasions, in some degree oblite-
rated from Jenner's mind the painful disappointment

which he experienced when he assayed to have his
first treatise on the Variolæ Vaccinæ published in
the Transactions of the Royal Society. He has
often mentioned the incident to me; but I wished
not to place it in a prominent situation, without at
the same time showing that the person who at first
was swayed by prejudice did not continue in error.
I know not whether Sir Joseph had any secret ad-
visers,* or whether, judging as others had done, he
thought the matter altogether so extraordinary, that
it could not with propriety find a place in the work
for which it was originally destined. Dr. Haygarth,
a most competent judge, entertained such sentiments
when the subject was first mentioned to him. We
need not wonder, therefore, that others might decide in the same manner. I fear, notwithstanding
every allowance, it must be admitted, that matters
were managed somewhat uncourteously. When the
subject was laid before to the President, Jenner was
given to understand, that he should be cautious and
prudent; that he had already gained some credit
by his communications to the Royal Society, and
ought not to risk his reputation by presenting to
the learned body anything which appeared so much
at variance with established knowledge, and withal
so incredible.

Such were the opinions which were entertained of
his discovery. I really am not inclined to bear hard
upon the memory of the persons who so reasoned.

* This point is set at rest by a letter of Dr. Jenner to Mr.
Moore, which will be found in a subsequent page.

Certainly they had not examined the subject fully; they had not investigated the admirable and conclusive evidence, as it is given in the inquiry ; and if they had, possibly conviction would not have followed. But be this as it may, Jenner was deterred from presenting his paper to the Royal Society; and that body thereby lost the honour of placing among their most valuable records a contribution, which has done more for the relief of human misery than any work that man ever produced.

Towards the close of this year he was gratified with an offering from an Italian muse. " Il Trionfo della Vaccinia, Poema di Gioachino Ponta, Genovese," was published at Parma, in 1810. The copy presented to Jenner bears this inscription in the hand writing of the author. " Al chiarissimo Sigᵣ. Dottore il Sigᵣ. Edward Jenner, Genio benefico dell' umanità, in signo d'alta stima e devozione, l'Autore. Napoli, 9th Nov. 1811."

This work is beautifully printed, and with the notes occupies 286 pages. The author, in transmitting it, sent a very handsome letter to the following effect :

In the course of last spring I did myself the honour, after having presented it to my Mæcenas,* to send you a copy of my poem on the Triumph of Vaccination, for which mankind is so much indebted to you, and by which your name will be immortalized. The consul of Naples, who now resides at Tunis, was to have forwarded to the English consul this

* Joachim Napoleon, King of the two Sicilies, ill-fated Murat !

book, together with another copy, which I sent him as a present, to induce him to take all possible care to convey this to you. The ship which was conveying it was taken, and all the goods of the consul sequestered or stolen. It happened fortunately for me that your consul at Palermo knew of the arrest of these two volumes, and got them liberated. I do not know if he perused the letter which accompanied the copy dedicated to you; I had no farther knowledge of the issue; the consul only knew that yours was sent to London. I will fain hope that you will have received it; but should it not be so, in order to fulfil the great desire I have that you should receive it, I send you another copy through the means of the courteous Mr. Graham, a Scotchman, who being exchanged for another distinguished prisoner of war is returning to his own country.

I beg you, Sir, to accept this humble present of a work of which you are the sublime inspirer, and by which I have endeavoured to add to the rays of your glory. If peace should be concluded, and if I should have the means, which the post is now deprived of, I would myself present you with this tribute.

Accept in the mean-time of the great esteem and good wishes of my heart, and the highest sentiments of consideration from,

Sir,

Your obedient humble servant,

GIOACHINO PONTA.

Naples,
Nov. 9th, 1811.

This letter was translated, and a very faithful analysis of the poem itself was made from the original, by an excellent and accomplished lady, Miss Eliza Jenkins of Stone. She enjoyed the respect and friendship of Dr. Jenner, and most deservedly;

for her acquirements in ancient and modern literature, as well as her numerous virtues, fully entitled
her to the distinction. Stone is in the immediate
vicinity of Berkeley; and some of the happiest
hours of the latter period of his life were spent in
the society of this lady and her amiable sister. The
poem of Ponta is divided into six cantos, and seems
to have been formed on the ancient epic model.
The action, or fable, as well as the characters and
sentiments, evidently all have a reference to that
lofty style of composition.

The action centres in the discovery of the antidote
to the awful infliction of the small pox. The machinery employed by the author in bringing about
so great an event is such as modern poets have for
the most part shunned. Gods and goddesses, spirits
and invisible agents, play their parts; and the
Fates ultimately reveal the secret to the " Immortal
Jenner." After proposing the subject, and describ-
ing the sufferings inflicted by the *Arabian disease* (as
he styles it), and the insufficiency of all efforts to
mitigate them, the pity of the Gods, after long delay,
was excited, and Jenner was created to triumph over
the malady. Then follows a tribute to the genius
of the family of Napoleon. But it is not on martial
exploits that the author intends to invoke his muse;
but to give an account of the ravages of the pestilence
partly allegorical, and partly founded on the generally received opinion of its breaking out in the time
of the pseudo-prophet. The first three cantos describe the consternation of Mahomet, at seeing death

armed with such tremendous weapons. He invokes
all the people subject to his authority to offer a
grand holocaust to appease the offended deity. The
other cantos contain a strange mixture of real with
ideal personages,—Aaron, Æsculapius, Hygeia, a
Gnome (an evil genius),—all contribute to carry on
the action. Towards the conclusion of the fourth
canto, after introducing Jupiter, Venus, and Europa,
and sundry other personages, the action is completed, and it was blazoned forth, that the discovery
which was to counteract the effects of the pestilence
was to emanate from Berkeley, the birth-place of
the *immortal Jenner.* To facilitate this great object,
Jupiter decrees that Oromazes (a good genius) shall
descend to the earth, to counteract in some degree
the evil done by the Gnome, by imparting to Jenner
the great secret.

> Scendi in Glocestro con veloci piume,
> Ivi uom vedrai, che in fronte ha una fiammella,
> E quando il mondo del diurno lume
> Fia muto e sveli notte ogni sua stella,
> Tu sfolgora di luce oltre il costume,
> E di te mostra fa subita e bella,
> E di' a quel Sofo, Giove a te la cura
> Dà di far scevera del Vaiuol Natura.
>
> Canto quinto, p. 186.

The winged minister of Jove's high will flies to
the banks of the Severn, and finds Jenner engaged
in learned lucubrations on the Variolæ, and sees a
lovely band of virtues in his company. This divine

genius then announces that his good and merciful
actions had been seen, and that he was destined to
still greater.

> Allo splendor raffigurò Jennero,
> Che vigil stava il medico talento
> Affaticando in freddo esame austero
> Meraviglioso ad operar portento ;
> E appunto al morbo Arabico il pensiero
> Di gran calcoli cinto aveva intento,
> E vide, che compagna a Jenner era
> Delle Virtudi la piu bella schiera.
>
> Canto quinto, p. 188.

The sixth canto is chiefly occupied with the progress of the vaccine discovery ; and there is a pretty
and interesting account of the labours of the different distinguished persons who exerted themselves
in this cause. He lastly invokes the good genius to
protect it, and entreats the young of both sexes to
offer a due tribute of honour to Jenner.

> A Jenner questo altare, e il simulacro,
> E ai suoi seguaci e ai Re propizj è sacro.

This is not an improper occasion to mention
another effort of a foreign muse, entitled, " La
Decouverte de la Vaccine, Poëme en trois Chants,
par un Médecin." The author, who signs himself
P. Py, D. M. M. (Aude) Narbonne, 18 7bre. 1816,
discussed the origin of vaccination, the restoration of Louis XVIII. the merits of Jenner, and
the benefits of his discovery, with great fervour.

In this collocation he has assumed the licence of a poet; and while he gives to England the honour of the discovery, ascribes to " *la belle France*" the glory of protecting and disseminating it. Much space cannot be allowed for a long quotation, but the following sentiments respecting Jenner are true in themselves, and gratifying from the pen' of a French physician :—

C'est lui qui nous apprit, du fond de l'Angleterre,
Le remède au fléau des enfans de la terre ;
Fut il jamais pour l'homme un sort plus glorieux
Que celui qui la rende l'emule de nos Dieux ?
De l'aurore au couchant, de Tamise jusqu'a Gange,
Tout se doit a *Jenner*, pour chanter sa louange.
Seul, il a consacré notre destin futur ;
Seul il merite aussi notre encens le plus pur.

A gentleman, whose progress ·in literature Jenner had watched with the interest and anxiety of a friend, and who had already proved the possibility of combining poetic talent with antiquarian research, again strung his lyre in " An Ode to Hygeia on the Vaccine Inoculation." I here allude to the Rev. T. D. Fosbroke, the author of British Monachism, the Encyclopedia of Antiquities, and many other works.

The origin of Vaccinia is thus described.

EPODE.

She was in verdant valleys born, Vaccinia hight ;
A curious, wizard sage, who aye would pore

On wondrous things from glens obscure as night,
  The goodly wand'rer to his cavern bore :
Long was she coy; nor e'er would speak, but blush'd.
  Philosophy, to that wise sage long known,
By him for this invoked, most softly hush'd
  Her fears, and drew her strangest things to own :
" She sprung (she said) of Io and of Jove."
  Pregnant of her, (though nought is said in song,)
When Juno's jealous ire had doom'd her rove
  In foul disguise amidst the horned throng,—
The angry sire exclaim'd, " Be thou divine—
The power o'er fell Variola be thine !"

Another, who ranked high among the poets of the age, and who had a deep affection for the discoverer of vaccination, planned a poem upon this theme, which after long deliberation he had convinced himself " was capable in the highest degree of being poetically treated." He announced his intention to Jenner in a very impassioned letter; but it is uncertain whether it was ever carried into execution. If not, one cannot but regret that so distinguished a writer as Mr. Coleridge should not have accomplished his design.

> 7, Portland-place, Hammersmith,
>   near London, 27th Sept. 1811.

DEAR SIR,

I take the liberty of intruding on your time, first, to ask you where and in what publication I shall find the best and fullest history of the vaccine matter as preventive of the small-pox. I mean the year in which the thought first suggested itself to you, (and surely no honest heart would suspect me of the baseness of flattery if I had said, inspired

into you by the All-preserver, as a counterpoise to the crushing weight of this unexampled war) and the progress of its
realization to the present day. My motives are twofold :
first and principally, the time is now come when the Courier
(the paper of the widest circulation, and, as an evening
paper, both more read in the country, and read at more
leisure than the morning papers,) is open and prepared for a
series of essays on this subject; and the only painful
thought that will mingle with the pleasure with which I
shall write them, is, that it should at this day, and in this the
native country of the discoverer and the discovery, be even
expedient to write at all on the subject. My second motive
is more selfish. I have planned a poem on this theme,
which after long deliberation, I have convinced myself is
capable in the highest degree of being poetically treated,
according to our divine bard's own definition of poetry,
as ' *simple, sensuous,* (i. e. appealing to the senses,
by imagery, sweetness of sound, &c.) *and impassioned.*"—O,
dear sir ! how must every good and warm-hearted man
detest the habit of mouth panegyric and the fashion of
smooth falsehood, were it only for this,—that it throws a
damp on the honestest feelings of our nature when we speak
or write to or of those whom we do indeed revere and love,
and know that it is our *duty* to do so; those concerning
whom we feel as if they had lived centuries before our time
in the certainty that centuries after us all good and wise men
will so feel. This, this, dear sir, is true FAME as contra-
distinguished from the trifle, reputation ; the latter explains
itself, quod iste *putabat,* hic *putat,* one man's echo of
another man's fancy or supposition. The former is in truth
φημη, i. e. ὁ φασιν οι καλοκάγαθοι, through all ages, the united
suffrage of the Church of Philosophy, the fatum or verdict
unappealable. So only can we live and act exempt from the
tyranny of time : and thus live still, and still act upon us,
Hippocrates, Plato, Milton. And hence, too, while reputation in any other sense than as moral character is a bubble,

fame is a *worthy* object for the best men, and an awful duty to those, whom Providence has gifted with the power to acquire it. For it is, in truth, no other than benevolence extended beyond the grave, active virtue no longer cooped in between the cradle and the coffin. Excuse this overflow, and let me only add, that most grateful am I, and a consolation it is to me for my own almost uselessness, that what I could most have wished to have done,—yea, had in lazy indefinite reveries early dreamt about doing,—has been effected in my own lifetime, and by men whom I have seen, and many of whom I have called my friends; in short, that I have known and personally loved Clarkson, Davy, Dr. Andrew Bell, and Jenner.

But while I gratify my own feelings I am pressing painfully on yours. I will, therefore, avail myself of an accident to change the subject. A very amiable lady, a particular friend of mine, and dear to me as a sister, has been subject generally once or twice in a year to a severe tooth-ache. She has many decayed back teeth; so many, as to put extraction almost out of the question; and besides, from the circumstances of the case, and the manner in which her face, eyes, and head are affected, I am convinced that the locality of the pain is in a great measure accidental; that it is what I have heard called a nervous rheumatic affection, and possibly dependent on some affection of the stomach or other parts of her inside. She is single, about six and twenty, has excellent health and spirits in all other respects, and bears this affliction with more than even feminine patience. Hot topical applications, such as tinctures of the pyrethrum, with ether, oil of cloves, &c. &c. give only momentary relief, or rather palliation. Her last attack was in November last, when she was confined to her bed more than a month by it, and reduced to a skeleton. Yesterday she had a return, and I am sadly afraid of another fit of it. Should you remember any case in point in the course of your practice, and be able to suggest any mode of treatment,

I will not say that I should be most thankful, but only that you will make a truly estimable family both grateful and happy. My friend, Mr. Morgan, has under Mr. Andrews, to whom you were so kind as to give him a letter of introduction, got rid entirely of his complaint. Though I still suspect it to have been symptomatic of some *tendency*, at least, to schirrus in some of the viscera, the liver probably. I have somewhere read or heard, that ipecacuanha in very large doses, so as not to act as an emetic, but as a sudorific, has effected great cures in rheumatic affections of uncertain, and, as they say, *nervous* kind. I have not yet read the answers, &c. &c. of Davy and Murray on the oxymuriatic, whether a chemical element or a compound; but I own, that in Davy's first communication to the R. S. I appeared to myself to see a laxer logic than is common with him. I judge merely as a logician, taking the facts for granted, and applying the rules of logic as an algebraist, his rules to X. Y. Z.—With every wish for your life and health, believe me, dear sir,

Most sincerely your respectful
friend and servant,
S. T. COLERIDGE.

Be pleased to remember me to Mr. Pruen should you see him.

I subjoin one more letter. It is from the Honourable and gallant Admiral Berkeley, who was then in command in the Tagus. The period at which it was written, as well as the events to which it refers, and the cordial friendship which it expresses for Jenner, all conspire to give it interest.

Lisbon, March 16th, 1812.

MY DEAR SIR,

Many thanks for the vaccine matter, as well as for your letter of December 1st, which I only received yesterday, owing to the detention of the Melpomene at Spithead. We have had a very cold winter, for this climate, with more frost than has been known for a long period, and, even now, the weather resembles more the month of March in England than Lisbon. Your anecdote of the gods who squabbled about hot or cold baths amused me. And I hope to dedicate Cullum's new ones to Hygeia, and that the Portuguese goddesses will dip their charms with more convenience as well as delicacy. I have procured an old fort by the waterside, which was formerly used as a custom-house, and there being a very good house in it, Cullum has fitted it up for warm as well as cold baths, and a very convenient as well as an elegant speculation it will be. No such thing in Lisbon, and very much wanted. It is astonishing how much more the rheumatism affects every body here than in our climate. The air is so much thinner and sharp, and the transitions from heat and cold more frequent. Your account of the *eau medicinal* has stamped a credit upon it with me, that I own I did not give it before, and I should not hesitate taking it, if a paroxysm of gout should occur. But I think the rheumatism has driven him out of the house. Our troops are, in general, in the most perfect health : but it is surprising how the ague has been felt amongst the officers, and how it adheres to them, like the Walcheren fever. My son has, thank God! enjoyed his health throughout the whole campaign, and has seen the whole of it, with credit to himself and satisfaction to his superiors. God preserve him through the whole, for he is a credit to his name. I am going up myself to the siege of Badajos, by desire of Lord Wellington, to meet him, and to see the effect of a noble battering train, which I have furnished from the navy. Lady Emily desires to be most kindly remembered to you, and is tolerably well, although the complaint still exists in the shape of a dumb gout, which swells her joints, and is painful at times, without inflammation or redness. I believe Dr. Cullum intends thanking you for all your kindness and

civil inquiries, which Dr. M'Neil, whom you saw at Cheltenham some little time since, repeated to him this day.
Of our native town you may conceive my affectionate regard
for it to be equal to yours. While I have strength to hold
my pen, I will sign myself your ever sincere friend, and
very faithful obedient servant,

G. BERKELEY.

CHAPTER VI.

LIBELS—PROGRESS OF VACCINATION—DEGREE OF M.D. BY
DIPLOMA CONFERRED UPON DR. JENNER BY THE UNI-
VERSITY OF OXFORD — DISCUSSIONS IN THE COLLEGE OF
PHYSICIANS — LORD BORINGDON'S BILL TO RESTRAIN SMALL-
POX—JENNER'S OCCUPATIONS IN THE COUNTRY—HIS IN-
TERVIEWS WITH THE DUCHESS OF OLDENBURGH AND THE
ALLIED SOVEREIGNS — ADDRESS FROM BRUNN IN MORAVIA
—LETTER OF SŒMMERING, WITH DIPLOMA, FROM MUNICH.

AFTER an anxious and more than usually protracted residence at Cheltenham, I find Jenner at
Berkeley, early in February, 1812. Independent of
domestic causes of disquiet, he was particularly an-
noyed by the atrocious falsehoods of the anti-vaccinists. Some of his friends were inclined to urge him
to seek redress in a court of law; but I am most
thankful to say that he did not follow their counsel.
He had better advice conveyed to him by Mr. (afterwards Baron) Garrow, who spoke on the occasion like
a man of wisdom, and with a full perception of the
real weight of Jenner's character. When the subject was mentioned to him by a friend, he observed,

" the truth probably is, that I should have been too
vain to have been consulted by the greatest human
benefactor of the human race, and am therefore mor-
tified by any thing which may deprive me of that
gratification of my pride and vanity. You may,
however, tell him, that there could indeed be very
few libels which I should suffer such a man to dignify by noticing." Never was sounder advice given;
for, I believe, the libels and the libellers are both
forgotten.

The affairs of the National Vaccine Establishment
were conducted with vigour and effect under the
superintendence of the new board, of which Sir
Francis Milman was the head. Mr. Moore, who admired Jenner for his talents, and loved him for his
virtues, executed the duties of his department with
unremitting zeal. He continued to lament sincerely
that Jenner was separated from the establishment.

The Report for 1812 was of a peculiarly gratifying
nature. Many of the most important documents
which it contained were supplied by Dr. Jenner. A
few of these are of so striking a nature that I must
mention them.

At the Havannah, though the small-pox had been
extremely fatal in that city, no death had occurred
from that disease for two years. In the Caraccas
and in Spanish America, the small-pox had been extinguished by vaccination. The same beneficial results were obtained both at Milan and Vienna, in which
latter place the average mortality from small-pox had
amounted to eight hundred annually. The returns
from France have been already alluded to in the first

volume. From a report presented to the class of Physical Sciences of the Institute by Messrs. Berthollet, Percè, and Halle, it appears that of 2,671,662 persons properly vaccinated, only seven had taken the small-pox, which is as 1 to 381,666. It is added that the well-authenticated instances of small-pox after small-pox are proportionably far more numerous.

In Russia, from the year 1804 to 1812, 1,235,597 were vaccinated, and in two years (1810 and 1811) the returns from the different stations in the Presidency of Madras gave 305.676 vaccinations; of this number 245,125 were Hindoos. The following important and interesting document accompanied the Russian Report. It is from the pen of Sir Alexander Crichton.

St. Petersburgh, 12th Sept. 1812.

DEAR SIR,

The reestablishment of peace between England and Russia being happily concluded, I embrace an early opportunity of sending you a letter on the state of vaccination in this empire, as I am convinced that the encouragement it meets with from the government, its gradual extension and success, cannot fail to be interesting to you. As all the reports on this subject, and indeed all those which regard every branch of medical police, or the health of the inhabitants, are addressed to me, it is in my power to give you the most accurate acco nt of the progress of your beneficial discovery. Having been, as you well know, one of your earliest advocates in England, and having never wavered in my opinion concerning its great advantages, you need not doubt that I do all in my power to encourage and support vaccination in Russia.

I have annexed for you a list which I caused to be made
out, of all the children who have been vaccinated in the Russian empire from the year 1804 to 1812. It is arranged in
such a manner that you will see the progress of vaccination
in each government during each of the intervening years
between 1804 and 1812, and to shew you the manner in
which the evidence is collected, I have added one of the half
yearly lists of one of the governments, translated by one of
my secretaries into French. I have chosen one of the
most distant and least civilised governments (Irkutsk) for
this purpose. It is inhabited, as you well know, by different
tribes of Tartars, chiefly by the Bouriates (or, as in England
they are commonly called, Bourations), and by the Tungusians. We have also in this government a great number of
Mantchu Tartars, being probably the original stock of the
Mogul race. You will find that one of your most zealous
inoculators in these distant regions is a priest of the great
Lama, himself a Lama.

The whole number of children inoculated in the empire,
concerning whom the government has received certain intelligence, amounts to 1,235,597. Now supposing, according
to a well-founded rule of calculation, that before the introduction of vaccination every seventh child died annually of
the small-pox, vaccination has saved the lives in this empire
of 176,514 children; and in an empire like this, where the
population is a great deal too scanty in proportion to its extent, such a saving of human life is of the greatest importance. Many generations must pass away before Russia
will have any occasion to dread Mr. Malthus's predictions.

In May 1811, the Emperor signed an ukase which has
given more activity to vaccination throughout his empire
than formerly existed.

ORDERS.

1. All the clergy to co-operate with the beneficent views
of the Emperor in destroying the prejudices which exist

among the people against the inoculation of the cow-pox,
or as it is now called in Russia the pock of surety.

2. To establish in each of the capitals and in every government town a committee of vaccination, consisting of the governor, vice-governor, the marshal of the nobility, the most
distinguished clergyman and the first physician of the
government. This Committee issues its orders and advice
to committees established in each district of each government,
and those inferior committees are composed of the mayor of
the district town, the first clergyman of the place, and the
physician of the district, and the *intendant* of the district.

3. The duty of the committees is,

(a) To keep exact lists of all the children in the government who have not been inoculated.

(b) To see that all children be inoculated by proper
people.

(c) To furnish good matter and proper instruments for
vaccination.

(d) To cause instructions to be given to pupils who devote
themselves to this occupation.

4. Permitting the committees to enact such by-laws, and
to adopt such modifications of measures as local circumstances may require. They are further ordered to make out
accurate lists of all who have been inoculated in the districts
and consequently in the government, and to transmit them
to the minister of the police, from whom they are sent to
my office.

5. The Committees of vaccination are ordered to see that
the practice and the art of vaccination be introduced into all
schools and seminaries, and that the students of all classes
be able to practise it before they leave the seminaries, and to
see that all midwives be properly iustructed in this art.
The said Committees are to distribute a popular work on
vaccination, printed at the expense of the crown, in all the
languages in use throughout the empire, and which contains
a clear, but abridged history of the disease, its real signs and

manner of distinguishing the spurious kinds, with rules when to take the matter to inoculate, and treat the inflammation when accidentally increased, &c.

Three years are allowed for vaccinating the whole empire, after which period there must not be found man, woman, or child, the newly born excepted, who have not been vaccinated.

The same ukase established appropriate rewards for the members of committees, and for inoculators who have been zealous in this good work.

I have sent you, along with this letter, a case of instruments, which I have approved for vaccination, and which is distributed gratis to the committees. 1 have also sent you some of the caricature prints in favour of vaccination. These operate as much on the minds of the poor peasants as the most eloquent discourses of the clergy.

Notwithstanding the supreme order of His Imperial Majesty, that all his subjects be vaccinated within three years, we find, that powerful as His Majesty is, this cannot be executed. There is a power greater than sovereignty, namely, the conscience or religious opinions of men, and in one or two of the distant governments there exists a peculiar religious sect, belonging to the Greek church, who esteem it a damnable crime to encourage the propagation of any disease, or to employ any doctors, or to swallow any medicines under the visitations of God. Reason has been employed in vain with these poor people; they have been threatened with severe punishment in case they remained refractory on such points, but all to no purpose. They have no priesthood, but attempts have been made to gain those of the community who have most influence with them, but all to no purpose. You may well imagine that no punishment has been employed, though threatened; and the government has come to the wise resolution of leaving this dispute to time.

I have thought that this short account of the state of vaccination in Russia would be acceptable to you. You may communicate it to my old friend and acquaintance, Bradley, if you deem it sufficiently interesting for his Journal.

Dear Sir, yours very sincerely,

ALEX. CRICHTON.

The documents which were transmitted to Dr. Jenner from the authorities in Bombay and Bengal proved that in the territories under their jurisdiction vaccination was carried on with undiminished success. Measures likewise had been taken to introduce the cow-pox into the territories of the Rajah of Coorg and the island of Java.

When this island fell under the dominion of the English (in 1811), the late lamented Sir Stamford Raffles was appointed Lieutenant-Governor. With the wisdom and benevolence which distinguished all his public acts, he immediately took measures to diffuse and perpetuate the practice of vaccination. In adjusting the revenue of the country portions of land attached to each village were set apart for the support of a vaccine establishment; and a certain number of vaccinators were appointed for each district. These vaccinators were under the immediate superintendance of European surgeons, and the lands or Sawahs have had the word Jennerian attached to them. The SAWAHS JENNERIAN thus handed over in perpetuity to support the cause of vaccination are recognized in all the rent rolls of the country.

Sir Stamford Raffles on his arrival in England in 1816, communicated these facts to Dr. Coley of Cheltenham, who published an account of them in the Medico-Chirurgical Journal for February 1817.

He also transmitted them to Dr. Jenner, who acknowledged his kindness in the following letter.

TO DR. COLEY, CHELTENHAM.

MY DEAR SIR,   *Berkeley, Sept. 5th,* 1816.

I am greatly obliged to you for your communication respecting the introduction of the vaccine into the island of Java, and beg you to present my respectful compliments and thanks to Mr. Raffles for his very interesting letter on the subject, and to assure him how happy I should be in having the honour of a visit from him at Berkeley. I certainly would pay my respects to him at Cheltenham, were I not at present so entangled with a variety of engagements.

It would doubtless be gladly received if a copy of the governor's letter were sent to the National Vaccine Establishment, which is too complimentary for me to think of sending myself. It strikes me that it would prove beneficial, if some copies of the instructions for conducting the process of vaccination were sent from the National Vaccine Establishment to Java, as they contain some minutiæ perhaps not yet known to the medical practitioners there. Some of the Annual Reports of the Establishment would also be read with interest, and I am certain that Dr. Hervey would be glad to send them.

I cannot conclude without thanking you for your laborious exertions during the late epidemic small-pox at Cheltenham ; but how shocking it is to think, that the labours of any medical man should be called forth *at the present period* on such an occasion; and in a town where, for a long series of years, I daily offered my services gratuitously to the public. By your saying nothing on the subject, I infer that the small-pox has now quitted the town.

Believe me, dear Sir,

very truly yours,

EDWARD JENNER.

Sir Stamford accepted Dr. Jenner's invitation, and

visited him at Berkeley. It may well be supposed that two such persons could not meet without finding many kindred themes for mutual discussion and edification. Sir Stamford's pursuits in natural history had been carried on with unbounded energy, and on the most magnificent scale. He had not then, it is true, made that splendid collection, which promised so much to enrich our knowledge of the productions of the East, and the destruction of which he was doomed to witness. The labours of twenty years all swept away, reduced to ashes in a few hours, was a trial which few men could have borne with equanimity.

The University of Oxford at the commencement of vaccination had been injudiciously importuned to confer an honorary degree on Jenner. This attempt was very properly resisted. Such a common and undistinguishing mark of academical approbation was held not to be sufficient if the discovery turned out to be as important as it promised ; while, on the other hand, it did not become such a body to move in a matter of this kind till the value of the discovery had been fully proved.

In the year 1813 the question was brought before the university, and in full convocation the degree of M.D. by diploma was unanimously voted to him. This proceeding, not less honourable to that distinguished body than to the individual who was thus signalised, it was imagined would open the portals of the College of Physicians to him, and remove all objections to his taking a seat at the vaccine board ; but it will appear in the sequel that it turned out otherwise.

He invited me to accompany him when he went
to receive the diploma; and I did so with much
satisfaction. We left Cheltenham on the morning
of Tuesday the 14th of December 1813, and arrived
at Oxford in the evening. He was playful and ingenious as usual, during the progress of our journey,
but at times a little depressed by anxiety for his son
Robert, who had just returned from school with
cough. He said that he so much resembled his son
who died, that he could not but feel alarm.

Next morning he was waited upon by Sir Chris-
topher Pegge and Dr. Kidd, the professors of Anatomy and Chemistry. They presented the diploma
with becoming expressions of respect. They mentioned that it was an honour which had not been conferred on any man for nearly seventy years before.
Jenner behaved with much simplicity and dignity.
" It is remarkable," said he, " that I should have
been the only one of a long line of ancestors and
relations who was not educated at Oxford. They
were determined to turn me into the meadows, instead of allowing me to flourish in the groves of
Academus. It is better, perhaps," he then observed, " as it is, especially as I have arrived at your
highest honours, without complying with your ordinary rules of discipline." He then reluctantly put on
his gown and cap, because, he said, the thing was
unusual to him, and he could not help thinking that
he should be an object of remark in the eyes of
others, forgetting for the instant that he was at Oxford, and not at Cheltenham.

Almost all the learned societies in Europe had con-

ferred their highest honours on Jenner. In thus
coming forward to mark their sense of his eminent
services and merits, they truly thought that all steps
and gradations by which the fitness of ordinary men
for such distinctions are ascertained, might in his
case be dispensed with. There was no risk that such
relaxation of the rules of academical discipline could
ever lead to the establishment of an injurious prece-
dent. Claims like Jenner's ever have been of rare
occurrence, and the elevation of such a person, so
far from tending to depreciate the value of literary
or scientific distinctions, could have no other effect
than to give them additional influence and lustre
in the eyes of every just and generous man. The
University of Oxford, jealous though she be of her
honours, had not scrupled to give a noble example
to all other learned corporations. Unfortunately
this example was not followed, especially in one
quarter where it would have been most graceful and
becoming. Many members of the College of Physicians of London had strong and right feelings on
this subject. They were conscious that the name
of Jenner might well have been associated in fellowship with that of Linacre, Caius, or Harvey, and
gladly would they have accomplished this object,
though not in the usual course required by their
statutes. Several ineffectual efforts were made to
bring about this very desirable measure, even before,
I believe, the Oxford diploma had been granted.
This occurrence, it was conceived, would immediately
remove the strongest objection. "No," said some
of the fellows, "it is true that Dr. Jenner, coming

from Oxford as he does, may, if he chooses, claim admission into our body, but he can only take his place with us after undergoing the usual examinations." The individuals who thus reasoned, after protracted debates carried their point. From a letter about to be submitted to the reader, it will appear that Dr. Jenner was no party to any of these transactions. He never courted this nor any other such testimony ; and it may perhaps be questioned whether the learned body did not lose an opportunity of conferring as much honour upon themselves as they could have bestowed upon the author of vaccination. This, I am well assured, was the conviction of many of the most distinguished of its members, and no one entertained it more strongly than the late generous and warm-hearted Dr. Baillie. If I am not much misinformed, he spoke his sentiments with unusual animation and warmth. As soon as Dr. Jenner knew of the discussion that had taken place, he wrote to the following effect to Dr. Cooke of Gower Street.

You saw by my reply to your first letter that I was not ambitious of becoming a fellow of the College of Physicians ; your second has completely put an end to every feeling of the sort, and I hasten to request you to stop the progress of anything that may be preparing for my approach to Warwick Lane. In my youth I went through the ordinary course of a classical education, obtained a tolerable proficiency in the Latin language, and got a decent smattering of the Greek ; but the greater part of it has long since transmigrated into heads better suited for its cultivation. At my time of life to set about brushing up would be irksome to me beyond measure : I would not do it for a diadem. That indeed would be a bauble. I would not do it for John Hunter's Museum; and that, you will allow, is no trifle. How fortunate I have been in receiving your kind communication! If the thing had gone on, it would have been embarrassing to both parties. I wish you would frame a by-law for admitting men among you who would communicate new discoveries for the improvement of the practice of physic. On this score (not alluding to vaccination), I could face your inquisition with some degree of firmness.

March 15th, 1814.

The next important public event that was intended to bear upon the practice of vaccination was a legislative measure to modify and restrain within certain limits the practice of small-pox inoculation. It was impossible by a peremptory decree, such as had been issued by the autocrat of Russia, to enforce vaccination in this country; but it was perfectly consistent with our usages and constitution to place the practice of small-pox inoculation under some degree of restraint. In consequence of the adoption of vaccination by most respectable medical men, many of the lower classes took up the small-pox lancet, and disseminated the disease in a very frightful manner. Some medical men, too, surrendered their own judgment at the bidding of their patients, and did not scruple to employ small-pox, when required so to do, even though they preferred vaccination. This was a case of conscience, which, as has been already stated, was warmly taken up in Jenner's own county; and those who did so were desirous that their reasoning should be admitted and acted upon by every professional man in the kingdom. They felt that the fears of the public were likely to be fatally cherished by the lamentable indecision of every man who could be induced for any consideration to vaccinate with one hand, and to variolate with the other. The absolute abandonment of the latter practice by all who had satisfied themselves of the security afforded by vaccination, was considered to be a duty of a plain and commanding nature. There was great difficulty in getting some men to adopt this view of the question, and a few rather influential persons thought it not inconsistent publicly to recommend cow-pox, and yet to employ small-pox inoculation when required to do so.

I am happy to say, that two years afterwards our principle was followed by the Colleges of Surgeons, both of London and Dublin, who publicly pledged themselves not to inoculate with small-pox ; and they recommended all their members to enter into similar engagements.

With all these convincing testimonies from learned men, and unquestionable proof of the power of cow-pox in controlling and extinguishing small-pox wherever it was generally practised, it is rather humiliating to find Dr. Jenner traduced and libelled, and small-pox itself diffused in such an alarming and unrestrained manner, that it was attempted to check the latter evil by a parliamentary enactment. The board of the National Vaccine Establishment, I believe, not only approved of this measure ; but, if I am not misinformed, the heads of the bill were digested and arranged under their superintendence.

The bill itself for regulating the practice of small-

pox inoculation, and checking the diffusion of that
disease, was brought before the House of Lords on
the 1st of July by Lord Boringdon. His lordship,
on introducing it to the House, took occasion to observe, that the principle on which it was founded
had often been acted upon and recognised by the
legislature. He showed, that in all civilised commu-
nities, individuals were restrained from exercising
unlimited dominion over their persons or property ; and that many statutes had been passed for
preventing the spreading of contagious diseases. He
particularly dwelt upon those which had been enacted
in order to stay the progress of the plague. Arguing
on these premises, he rightly contended that similar
provisions ought to be put in force against the plague
of small-pox. He therefore recommended that regulations should be adopted in all cases where this disease existed, either in consequence of inoculation or
casual infection. His lordship's design was not to
interfere with the will of parents in adopting either
vaccination or small-pox inoculation. " If, however," said he, " the latter be chosen, it is desirable
that the surrounding neighbourhood should as much
as possible be protected from the spreading of the
contagion." To secure this very desirable object,
the bill contained sundry clauses and enactments,
which I am sorry to say are still much called for.
Others were of a more questionable nature, and were
rather calculated (in this country, at least,) to give
an air of levity to a very serious subject. In the
case of children inoculated at the expense of the
parish, it was very properly intended that the vaccine

should alone be employed. The bill was opposed by the Lord Chancellor, and by Lord Ellenborough. Both these noble lords contended that the common law as it now stands was better calculated to prevent the spreading of the small-pox than any of the provisions of the present bill. Lord Ellenborough, after clearly and forcibly explaining the common law, and ridiculing some of the clauses of Lord Boringdon's bill, is reported thus to have spoken of vaccination itself. "No doubt," he observed, "it was of some use, but he did not concur in all the praise bestowed upon it in this bill; but if the noble lord considered it a complete preventive of the small-pox, he differed with him in opinion. At the same time, he had shewn his respect for the discovery, for he had had eight children vaccinated. He believed in its efficacy to a certain extent: it might prevent the disorder for eight or nine years, and was desirable in a large city like this, and where there was a large family of children."

After a few words from Lord Redesdale, and a short reply from Lord Boringdon, the bill was withdrawn. It was certainly rather a crude measure; the enactments had not been maturely considered; some of them were inconsistent with the habits of the country, others were impracticable; and the solid principle, alike recognised by the common and statute law, was in some degree overlaid by the comparatively unimportant machinery of the bill. If, keeping steadily in view the principle just alluded to, it had been further enacted, that no one should be permitted to propagate small-pox by inoculation but a properly qualified professional man, and had the laws which forbid the exposure of persons labouring under contagious diseases been more clearly explained and enforced, every benefit that could have been expected from legislative interference in this country would have been obtained. The only new enactment that was required, is at once so simple and so just, that one can scarcely anticipate any opposition to it; and even now, were it adopted, it would be of infinite service.

The fate of the bill was, I believe, not unexpected by Dr. Jenner, but the concluding part of the observations of Lord Ellenborough were not less unlooked for than incorrect in themselves, and injurious in their consequences. I have seldom seen Jenner more disturbed than he was by this occurrence, and not certainly because he had any fears that the unsupported assertion of his lordship would prove correct, but because it unhappily accorded with popular prejudices, and when uttered by such a person, in such an assembly, was calculated to do unspeakable mischief. His friends took the same view of the subject, and were of opinion that something should be done to prove to the Chief Justice, that though his law might be good, his physic was bad. " Ponebat enim rumores ante salutem!" But how to convince so dignified a person, speaking from his privileged station in the Upper House, that this really was the case; that he was unsettling the confidence of numberless anxious parents; and that by attempting to deprive vaccination of more than half its virtues, he was promoting the practice which he professed himself will-

ing to control, was a question somewhat difficult of
solution.

It is a remarkable thing, that all evidence which
had been accumulated on the subject was in direct
opposition to his lordship's opinion. That derived
from casual cases of cow-pox carried us back to a
period of fifty or sixty years, and the direct testimony from the history of vaccination itself, led
clearly to the conclusion, that the protection which
it afforded against subsequent attacks of small-pox
was as great as small-pox itself gives. The declarations, however, of the Chief Justice too much
accorded with popular prejudices, and the anti-vaccinists were doing all in their power to make them
effective ; and doubtless were not a little proud of
the co-operation of his lordship. Jenner's own feelings on the subject were, I believe, somewhat excited by the incident which will be found in another
part of this volume. The noble lord who had very
indiscreetly put forward an unfounded statement in
Jenner's hearing at St. James's, and had been rebuked for so doing, might, it was supposed, have not
forgotten this occurrence when descanting on the
same subject in another place. The reader may
easily make allowance for the sensitiveness of Dr.
Jenner on the subject ; for, though he was the last
man in the world to be influenced by mere personal
considerations, it was not to be supposed that he
should be unmoved by the reiteration of sentiments
deeply affecting the cause of vaccination, but especially when uttered by an individual so high in station and authority. He therefore felt that the fallacy

of the reasoning should be exposed, and the assumptions on which it was founded contradicted by well attested and conclusive facts. A small pamphlet, containing such materials, was prepared by myself; but some delay having taken place in the publication, the subject became less urgent, and it was allowed to drop.

I now gladly turn to scenes of a different kind—scenes more congenial to Jenner, more in unison with the prevailing tone of his character. The atmosphere of a large city never accorded well with his taste : still less could he endure the strife engendered by professional jealousy or emulation. He turned away from them with aversion; and whenever he was compelled to encounter any thing of this kind, he resumed his peaceful and rural life with increased relish. I never saw this more strikingly exemplified than at this period. He had then occasional opportunities of breaking loose from all the sources of his care, and he enjoyed himself amidst the beauties of his neighbourhood in a manner which spoke as much for his wisdom, as it did for the purity and simplicity of his mind.

One of his favourite haunts was Barrow Hill. It is situated in a peninsula, formed by the first and boldest sweep of the Severn. It rises very little above the surrounding country, and the stranger on approaching it could form no conception of the nature of the scenery it commands. The road from the village of Frampton passes through a low, but rich and well-wooded part of the vale; but it promises nothing in the way of picturesque beauty or grandeur. The ascent from the bottom of the hill to the top is a beautiful green sloping bank, and the greatest elevation is very inconsiderable. It was Jenner's habit to seek the summit when the sun was going down, and when the rapid course of the sweeping tide had brought the Severn to its highest level. Under such circumstances, the view, from whatever point it was contemplated, was full of loveliness. Looking southwards, the river expanded to a great breadth, and in consequence of the situation of a prominent headland on either side, it assumed the character of a lake spread beneath the spectator's feet, bounded on one side by the bold and picturesque features of the forest of Dean, and on the other seeming to lose itself in the flat surface of the vale, which, after a deep woody expanse, is itself terminated by the hills. In the distance beyond this land-locked reach, the opening into the Bristol Channel bursts upon the eye, and its vicinity to the mighty ocean may be discerned by the character of the vessels which it bears upon its surface. Not like the little barks which lie in wait for the filling of the river to enable them to pass over its shoals and its sand-banks, a tall and stately ship may now and then be seen stretching across the estuary, and bearing her cargo, it may be, from the Sister Isle, or " Nations besides from all the quartered winds."

At the close of a summer's day, when a flood of golden light rushes through the openings which are formed by the deep valleys in the forest of Dean, and is reflected from the smooth surface of the water, the tranquil splendour of the scene is very captivating. The lazy vessels which had been lying with their sails furled now begin to give signs of life and motion. The anchor is weighed, and the wings are expanded to the evening breeze, which, aided by the swift and high-swelling tide, carries them proudly along, till they approach almost to the spectator, and sweep round the beautiful bend of the river that nearly encircles the spot on which he stands.

Looking westward and northward, the view is bounded by the forest of Dean. At one part it rises gently from the river side; at another, it seems to start boldly and abruptly, and to oppose a firm and strong barrier to the farther encroachments of the Severn. The church of Awre stands on a sloping verge; beyond it, rise in beautiful succession undulating banks covered with orchards and oak-timber. Close to the banks of the river, and at the bottom of the more abrupt rising of the forest, lies the town of Newnham, and immediately above it, on the brow of the hill, is the village of Little Dean. Carrying the eye along the reach of the Severn, from Newnham, a bold precipitous bank is seen: it finely contrasts with the surrounding objects, and shews its bare and almost perpendicular red surface where the contiguous landscape is soft and beautiful. This is Westbury Cliff, and here is the bone-bank, which has been already spoken of as one of the haunts of Jenner. He had a particular fondness for this scenery. It was endeared to him by many interesting associations. It was very near the residence of some of his oldest friends. It commanded a view of the trees which overshadowed his much-loved home; and he could also see Pyrton, Awre, the Hock Crib, Westbury Cliff, all interesting to him from the many and happy opportunities which they had afforded him for pursuing his favourite studies. In the summers of the years 1812, 13, and 14, he enjoyed several days of delightful recreation in re-visiting this spot. During the periods just mentioned he had, I believe, greater freedom from professional care, and was permitted more tranquilly to meditate on the great blessings which his discovery was daily conferring on mankind, than at any former period since its promulgation. Instead of courting the elevation which it naturally gave, he returned, with a mind uninfluenced by the distinction he had acquired, to all the simple enjoyments of early days.

He promoted a little social meeting, which he called the Barrow Hill Club. The members were few, consisting of himself, Mr. Gardner,* Mr. Henry

* Poor Gardner wrote a Poem, entitled Barrow Hill, descriptive of its scenery, from which the following lines are extracted.

> Whilst warm the vivid feelings glow,
> Softly I'll mount yon gentle summit's brow,
> And gaze below where rough Sabrina pours
> Along the bending vale her Cambrian stores.
> Charm'd with the scene, the stream prolongs its stay,
> And gently ling'ring winds its lengthen'd way.
> Triumphant view! New worlds before us rise,
> Flush on the gaze, and strain the busy eyes:
> The far-off blue-hill where the mantling cloud
> Weaves round its lofty brow a misty shroud;

Hicks, the Rev. Robert Halifax, myself, and now and then an occasional visitor. Two or three times in each of the summers just mentioned, we were in the habit of meeting at the Bell Inn in Frampton.

> The wide-stretch'd field with yellow harvests warm,
> The whiten'd cottage, the encircled farm :
> The stream's bold reach along its redd'ning side,
> The march majestic of its solemn tide ;
> The shades of rugged rock and fringing wood,
> That undulating dance adown the flood ;
> Queen of the western streams, hail Severn, hail !
> Wide boast of Cambria ! glory of the vale !

As Mr. Gardner's prose is better than his poetry, and as the following letter alludes to events quite in harmony with those described in the text, I subjoin it.

To Dr. Jenner.

Frampton, 21st May, 1817.

My dear Friend,

I received yours, with the vaccine virus inclosed, by Mr Pearson; I very much thank you for your kindness. I have long practised the method of inoculation you describe, agreeable to your former instructions. From some unaccountable causes, the fame of vaccination seems to decline in this part of the country : I find my offers of gratuitous service very frequently rejected even by those whose former children have undergone the operation.

Some parts of your letter awakened in my bosom emotions of the purest delight ; my mind was thrown back to the days of former years—the bread and cheese dinner under the fossil rock, the excursion to the garden cliff, and the dinner at Framilode passage ; and then, by a rapturous association of ideas, I was transplanted to the gardens of Berkeley, when in a sunny summer's morning we have traversed the town in search of

After a temperate meal there, we generally retired
to the hill. On such occasions Jenner's manner was
peculiarly attractive. He was as free from all pride
as though his name had never been heard of beyond

the most beautiful tulip, polyanthus, or carnation—your honoured departed relative being the chief of the party. But the
recollection is too much—the tear flows as I write.

My friend Mr. D. Loyd has lately presented me with the
works of the Ayrshire bard, Burns, which I never had before
seen, though many scattered pieces have occurred to my notice
in reviews and magazines. He is a poet of the first order, and
very superior to Walter Scott or Lord Byron, the fashionable
favourites of the day.

But what struck me very forcibly, was the masterly piece of
biography, written by Dr. Currie, which occupies the whole
of the first volume. I never saw so fine a performance of the
kind: it unites the variegated expanded eloquence of Burke
with the philosophical precision of Hume; it is equally conspicuous for just taste and sound criticism. He has aptly defined the nature of poetic talent, and truly appreciated the
character of its possessor; and although I am one of the lowest
of the tribe, yet I know I have some faint sparks of its fire;
and I feel, very sensibly feel, the truth of Dr. Currie's observations. I have felt it through my whole pilgrimage through
this wilderness world; yet I know not how it is, I feel at sixty
much less indolence than I did when I was thirty.

Currie far exceeds Gregory as a biographer: the latter's Life
of Chatterton is comparatively a poor, a very poor performance.
I should suppose that the author of the Memoirs of the Scottish
Bard must be a good physician; he certainly possesses the
chief requisite, " a vigorous, comprehensive, and philosophic
mind."

I will take the earliest opportunity of seeing you at Berkeley,
but it cannot be before the Midsummer holidays, which commence 20th of June. I can walk home with great ease.

Cannot Mr. Halifax, yourself, &c make a party at the Rock,

the limits of the district which the eye surveyed. His mind was full of alacrity, and the playfulness of his mirth most admirably blended with the instructive tone of his conversation, enlivened as it was by the beauty of the scene, and by early recollections.

The year 1814 was one of the most memorable in our memorable times. The abdication of Napoleon, the restoration of the Bourbon dynasty to France, the visit of the allied sovereigns to England, and the prospect of peace and repose to the world, gave reason to hope that the great ones of the earth, who, amidst strife and bloodshed, had expressed their obligations to the author of vaccination, would not forget him in the day of their triumph and of their glory. He had, in no small degree, indirectly aided the cause of their arms ; and something more than mere empty compliment was due to him for his signal services.

He had occasion to visit London in the end of April, 1814, and took up his residence at No. 7, Great Mary-le-bone Street. This, I believe, was the last visit he ever paid to the metropolis. He sojourned there for more than three months. His

similar to that of ours in the summer of 1782 (thirty-five years since! " Tempus fugit ! ") I can provide the bread and cheese and *fat ale*. This reverend gentleman is a truly generous and worthy character. He has kindly volunteered an annual deposit on my behalf, a favour unrequested by any one, and of which I never could have entertained an idea, since our acquaintance has been comparatively short.

I remain yours sincerely,

EDWARD GARDNER.

stay was somewhat prolonged in expectation of the arrival of the allied sovereigns. The Duchess of Oldenburg, sister to the Emperor of Russia, had had several interviews with him, and she was also present at a subsequent period when he was presented to the Emperor himself.

Jenner had a peculiar delicacy of perception in all that regarded the grace and dignity of the female character. His conversations with the duchess delighted him exceedingly, and made some amends for the impatience with which he endured a London life. Writing to myself on this subject, on the 18th of May, and urging me to pay him a visit there, which I did, he says, " Though I can't get away, yet I am quite sick of the life I lead here, and certain I am that your presence would relieve me. The mighty potentates will soon be here ; and, I suppose, I shall see some of them. The Duchess of Oldenburg is a more interesting being than I ever met with in a station so elevated." He concludes this letter with an incident of a domestic nature. " Poor Mrs. Jenner * has suffered severely during my absence. So ill was she, that I held myself in readiness daily to go down on the arrival of the post. My last accounts have been very pleasant. Judge what a life of disquietude I have here."

The account of his interview with the Emperor I insert very nearly in his own words. " I was very

* She was constantly attended by Mr. Wood, one of Dr Jenner's oldest and most attached friends. This gentleman enjoyed the doctor's confidence, and attended to all his affairs during his absence from Cheltenham.

graciously received, and was probably the first man
who had ever dared to contradict the autocrat. He
said, ' Dr. Jenner, your feelings must be delightful.
The consciousness of having so much benefited your
race must be a never-failing source of pleasure, and
I am happy to think that you have received the
thanks, the applause, and the gratitude of the world.'
I replied to His Majesty that my feelings were such
as he described, and that I had received the thanks
and the applause, but not the gratitude, of the world.
His face flushed ; he said no more, but my daring
seemed to give displeasure. In a short time, however, he forgot it, and gave me a trait of character
which shewed both great goodness of heart and
knowledge of human nature. My inquiries respecting lymphatic diseases, and tubercles, and pulmonary consumption had reached the ears of the Grand
Duchess. She was present, and requested me to detail to her brother, the Emperor, what I had formerly
said to her Imperial Highness. In the course of my
remarks I became embarrassed. She observed this,
and so did the Emperor. ' Dr. Jenner,' said she,
' you do not tell my brother what you have to say so
accurately as you told me. I excused myself by
saying that I was not accustomed to speak in such a
presence. His Majesty grasped me by the hand, and
held me for some time, not quitting me till my confidence was restored by this warm-hearted and kind
expression of his consideration." This circumstance
gave Dr. Jenner much satisfaction.

After the extraordinary events which led to the
expulsion of Napoleon, and when kings and princes

and warriors who had contributed to his downfall were about to visit the metropolis of that country which had invariably opposed his ambitious designs with uncompromising hostility,—it was thought that Dr. Jenner, who by his discovery had saved our own fleets and armies from the pestilential ravages of small-pox, and had rendered not less important service to the people, than to the military force of every potentate in Europe,—it was thought that such an individual, amidst the general shouts of triumph and of victory, would not be forgotten; that a congress of sovereigns, brought together by a most remarkable combination of events, would feel happy to signalize collectively, as they had done individually, an invention not the least wonderful of those which marked an era of wonders.

Dr. Jenner himself could not but feel a personal interest in such a combined and powerful display of general respect, should those, who on this memorable occasion were the representatives of the civilised world, think fit to shew it. The nature of his services was worthy a commemoration of this kind, and many thought that it would have been eagerly adopted. In this expectation, however, they were disappointed. Ingenuity was exhausted in heaping honours on those who had been engaged in taking away man's life, whilst the modest and unobtrusive individual, who could count millions preserved through his means, was permitted to enjoy his own self-satisfaction without any peculiar marks of public consideration.

Most of the distinguished foreigners were desirous

of being introduced to him.   In addition to the Emperor of Russia and his sister, he received a respectful intimation on the part of the King of Prussia, appointing a time for an interview.   His majesty was the first crowned head who adopted the practice of vaccination in his own family; Dr. Jenner having sent virus for the inoculation of the children of the princess royal of Prussia so early as the year 1799. The regulations which have been acted on in the Prussian dominions have been so efficacious as to give that country almost a complete immunity from small-pox.   He also had interviews with Blucher, Platoff, and most of the principal personages.

It is now, perhaps, of little moment to refer to occurrences which regarded merely temporal honours or advantages, and which have not any necessary connexion with Jenner's character or reputation. It is right, however, in attempting to delineate the history of any individual, to bring forward the events and transactions which made up the sum of his experience in passing through life.   It is on this principle that I mention the design that was entertained by some of his friends of addressing the allied sovereigns on his behalf.   Count Orloff, then Russian ambassador to our court, who had a great respect and friendship for him, took a different view of this subject, and suggested that Jenner himself should memorialize the assembled monarchs on the score of his claims as an universal benefactor.   He shrank from such a project, and it is better that it should have been so met by him than that he should have acted upon it, even though complete success might have been the result.

On looking back, even at the distance of a few years, on events which when they arose appeared all important, it is astonishing how their character changes if viewed in connexion with higher and more enduring objects. The smiles and the favours of earthly potentates might have been coveted for Jenner; he himself might have played the courtier on such an occasion, and he might have had his reward; but what would that have availed him? and what satisfaction would it give to those who now honour and revere his name had he been

"Stuck o'er with titles, or hung round with strings,"

if such outward distinctions had been purchased by a surrender of the smallest portion of his own personal integrity or dignity?

Another plan was at this time devised which had a different aim; and one which, I think, it would have been gratifying to have seen carried fully into effect. The mothers whose offspring had already been protected by the vaccine shield, the daughters whose beauty had been preserved by its benign influence, intended to unite in offering a tribute of their gratitude to the author of vaccination. To this object they were moved by the strongest feelings that can actuate the female breast. They remembered the anxieties and perils inseparably connected with the progress of small-pox, in whatever form it was communicated. Protracted suffering, hideous deformity, shocking to the patient, frightful to the beholder, often terminated by death, or by permanent injury to the individual, blasting a mother's joys or blighting her fondest hopes; these, and greater evils, more or less embittered the happiness of every family in the land before the adoption of vaccination. Mothers, seeing their offspring rescued from such a scourge; daughters, knowing that their beauty, their health, their life, had been preserved; and that, too, by the well-directed labours of one of the most meritorious and disinterested of their fellow-countrymen, did think that it became them to bear public and unequivocal testimony to such great and distinguished benefits. Such, at least, were the sentiments of some of the British fair. They were associated with many of their best and tenderest affections; and they were willing that the author of vaccination should not " fall away " till he experienced both their force and their sincerity. They were anxious to combine in one great and substantial effort the whole female influence of the realm, for the double purpose of cheering and rewarding the author, and of promoting the practice of vaccination.

This praiseworthy design promised, at one period, to be attended with considerable success. It was ardently promoted by many ladies, and a nobleman who has ever felt the deepest respect for Jenner, I mean Lord Segrave, applied to her late Majesty Queen Charlotte to patronize the scheme, to which she was graciously pleased to consent. It also found favour in the eyes of some of the highest personages in the kingdom; and there is great reason to believe that, but for the distresses of the country after a long and arduous warfare, it would have been crowned with success.

All these things passed away; and we find Jenner
again at the end of the summer of this year (1814)
in the midst of his domestic circle at Cheltenham.
Many events connected with the practice of vaccination itself kept him in a state of agitation; and his
family distresses were by no means slight. I had
witnessed many of them myself; and in a letter written to me at this period, he thus expresses himself:
" You know how actively I have been employed here
since my return from town; and the inexpressible
miseries I have endured from domestic affliction. The
three servants are still in bed, I think convalescent;
but there is no marked termination of the fever except in Frank, who suffers only from hunger, as
nothing seems to satisfy him. Mrs. Jenner is rather
better; but there is another on the sick list—alas!
myself. I was seized on Sunday with cholera, and
sad work it has made with me. Within these four
hours things have changed for the better, or I could
not have answered your letter by return of post."

Though the potentates did not, either in their individual capacity when in England, or when assembled
in congress at Vienna, confer any mark of distinction on the author of vaccination, some of their subjects did, nevertheless, bear him nobly in their memory, and testify their admiration in the manner
recorded in the following characteristic document.
I give it exactly as it was sent, convinced that all
will respect the feelings which incited the warm-
hearted inhabitants of Brünn to bring the historical
recollections of their country, the renowned deeds of
their ancestors, into close alliance with the honour

they were anxious to bestow on the inhabitant of another land, whose only claim to their veneration and esteem formed a striking contrast to the great events to which their letter alludes. The imperfect English, interspersed with half Latin and half German idioms, though it may raise a smile on the cheek of some fastidious critic, has added, in my mind, a deeper interest to the communication.

To the Right Honourable Physician Edward Jenner, Discoverer of the Cow-pock, the greatest Benefactor of Mankind.

At London.

MOST HONOURABLE DOCTOR,

At the most distant frontier of East Germany, in a country where the Romain's army two thousand years before triumphing, and 444 the savages Huns under the commande of Attila, and 791 the Emperor Charles, the Huns with success combatting, passed, and where the Swedes under Gustav the Great 1615 have made tremble the ground of the country by the thousands of cannons, and there where even 1740 the Prussians and 1805 the French warriors victorious appeared, in that remarkable country had the vaccined youth from Brünn, with the most cordial sentiments of gratitude to Thee, a constant monument with thine breast-piece in the 65th year of thine age erected, even in the same time as the great English nation, by her constancy and intrepidity, rendered the liberty of the whole Europa, and as the greats regents Alexandre and William passed through that country. Accept generously, great man, that feeble sign of veneration and gratitude; and Heaven may conserve your life to the most remote time; and every year, in the presence of many thousand habitants, a great feast near that temple is celebrated for the discovery of

vaccine. We will us estimate happy, if we can receive few
lignes to prove us the sure reception of that letter.

Most honourable Doctor,
yours most obliged servants,
Medicinæ Doctor RINCOLINI, physician.

CLAVIGER, { first surgeon and vacciner of
Vaccine Institute at Brünn.

Brünn in Moravia, the 20th *October.*

A drawing of the "*monument*," as it is called,
accompanied this letter. In the centre of the temple
the bust of Jenner stands upon a pedestal, on which
is the following inscription :

Divo Anglo
Eduardo Jenner,
LXV.
Ætatis ejus Anno
Vaccinata Brunensis
Juventus
MDCCCXIV.

I am not certain that "the vaccined youth at
Brünn" ever received Dr. Jenner's reply to this their
cordial and animated address. Should, however,
these pages reach their hands, they may be assured
that their kindness and their generous warmth in
addressing the discoverer of cow-pox was most
deeply felt and acknowledged by him. I was with
him when the document arrived ; and it gave
him unqualified gratification. I know that it was
his intention to have expressed such feelings for
himself; but as I am not equally certain that this
his wish was accomplished, it the more becomes me

to record his sentiments, lest any one should imagine that he was uninfluenced by an address which did, indeed, awaken more pleasing emotions than almost any other testimony he received.

Another communication, of a somewhat different description, I have great satisfaction in submitting to the reader. I regret that it did not arrest my attention when carrying on the discussion on the identity of small-pox and cow-pox in the first volume. The testimony of two such men as Sœmmering and Hoffman is worthy of all respect and consideration. I can scarcely conceive any thing more striking than the manner in which that testimony was given. It is not the less valuable from being founded on the very same observations which guided the decision of Jenner. Standing as it does, it claims respect and attention; but when supported, as it has been, by proofs of a different description, it carries with it the force of demonstration. Should any one take an interest in this question, they will not, I trust, find their trouble misapplied in reading the letter of Sœmmering in conjunction with the evidence contained in the fifth and other chapters of the preceding volume.

DEAR SIR, *Munich, Nov. 1st,* 1814.

I have the honour of presenting to you the diploma of our Royal Academy of Sciences, as a due acknowledgment of the superiority of that salutiferous genius, by whose infinite merit mankind stands delivered for ever from the most hideous and dreadful of all diseases.

Bavaria can boast of being the country in which your glorious discovery not only found the highest applause, but which from the very first beginning till the present day continued regularly and stedfastly its universal introduction.

As for my own part, having myself in my youth applied to study most minutely the appearances of the small-pox, and having examined it often through the microscope even after artificial injections of the cutaneous vessels, and recollecting the features of a certain sort of small-pox, and at the first look on the cow-pox was struck with its identity with the small-pox in their mildest—but, alas! by far rarest form; amongst others, I remember perfectly well of having observed, with my worthy friend the late Dr. Lehr, of Frankfort on the Mayn, particularly in two cases, after inoculation with the small-pox virus, on the arm of healthy children six weeks old, on the breast of their equally healthy mothers only one single small-pox without any other pustule besides any where else. This pearl-like pustule, surrounded with a fine red areola, had perfectly the same appearance as the cow-pox, kept such a mild and short course, that my friend doubted whether such an extraordinary slight inoculation could be sufficient to the purpose. His doubts were the reason why I was called by him in consultation in these cases. Our meritoriously most famous physician in Germany in regard to small-pox, Dr. Ch. L. Hoffman, Physician to the late Elector of Mayence, a lynx-eyed man, though more than eighty years old, regarding attentively the first cow-pox shewn to him, energetically exclaimed, " *This pox surely will secure against the small-pox, being indeed nothing else but a real and true genuine small-pox of the mildest sort ; and you all know that ten thousand poxes give no more security than a single one.*"

He used to tell me confidently, as a result of his long experience, " Believe me, friend, there exists a certain form or a particular sort of small-pox, so mild, so regular, and of so short a duration,—in short, of such benignity,—that the patient, whatever regimen he follows, this sort of small-pox by no means will kill him ; nay, even in any way hurt him."

But I must own I never saw a physician so extremely careful in the choice of the virus. He never lost a patient inoculated by himself under many thousands; and saw but one child marked, by the open fault of the mother.

Give me leave to add to these observations, perhaps long before known to you, that I invented and introduced the denomination, " guarding-poxes," now almost generally adopted, amongst other reasons amply detailed in my dissertation, " Prüfung der Schulz-oder Kuh blattern durch Gogen impfung mit kinder blattern."

I wished to denote by this denomination my idea of the perfect identity of the two diseases, " *morbos non sua natura sed gradu diversi.*"

May the blessings of so many millions whose lives you saved, or whose deformities you prevented, contribute to exhilirate the days of their benefactor. I am, dear Sir,

With the profoundest veneration,

Your obedient and humble servant,

DR. S. TH. VON SOEMMERING,

R. B. Gèheimer Rath.

The establishment of literary and scientific institutions in our provincial towns forms a striking feature in the character of the present times. The bearing and fructifying of the plant of knowledge, which, to use language that I love to quote, seems to have been appointed to this autumn of the world, was anxiously watched by Jenner. He even outran the spirit of the age, in attempting to cultivate this goodly tree. In this spirit he endeavoured to establish a literary and philosophical institution in Cheltenham in the year 1814. Several preliminary meetings were held at his house, No. 8, St. George's Place, in the end of the preceding year. On the 3rd of February 1814,

the first public meeting was held at the Assembly
Rooms, when Dr. Jenner was formally elected President of the Cheltenham Philosophical and Literary Society; and Mr. T. Morhall, Secretary. At
this time, the number of members had increased to
thirty. Papers were read by Dr. Parry (now of
Bath), Dr. Boisragon, and myself.

The Institution did not meet with that encouragement which had been anticipated. Shortly after its
establishment, Dr. Jenner's bereavement by the
death of his wife, and his consequent retirement
to Berkeley, put a stop to his exertions, and it soon
fell to the ground.

CHAPTER VII.

DEATH OF MRS. JENNER—EPIDEMIC AT EDINBURGH—PUBLI-
CATION OF SIR GILBERT BLANE'S TRACT—VARIOLOID EPI-
DEMICS.

A CALAMITY, which had manifestly been long suspended over Jenner, and which he had often looked
forward to with distressing apprehensions, was now
at hand. During the whole of the last year, Mrs.
Jenner's health, at all times feeble, became evidently
more impaired. In the spring, however, she seemed
to rally a little ; and the fears which were justly entertained for her safety, though not altogether removed, were in part mitigated. About the end of
August, in addition to her usual pulmonic symptoms,
she experienced an attack of bronchitis. Such a
seizure in such a person was more alarming than it
would have been under other circumstances. She
was so slender, so attenuated, and so much deprived
of all vigour of constitution by protracted illness,
that she could not have existed except under the
most constant care and vigilance. For many years

she had lived almost in an artificial climate ; and for a considerable time before her last attack she was confined entirely to her room.

I saw her very frequently at this time, and ha constant opportunities of witnessing both her husband's anxiety and her own patience and resignation. I visited her at Cheltenham the night before she expired (Sept. 13, 1815); and when she was in full expectation of the fatal event. The impression made upon my mind by the scene altogether I can never forget. She had long been preparing for her final account, and her departure was marked by those accompaniments which generally attend the death of the righteous.

I was with Dr. Jenner the day after this disastrous event : he grasped my hand with great emotion, and said, "Baron, I am a wretch !" Whoever has attended to the imperfect delineation of his character, which these pages contain, will readily perceive that feelings of this kind were not likely to be of a transient nature. His sensibility, though lively and acute, was capable of receiving the deepest and most abiding impressions. The death of Mrs. Jenner may be considered as the signal for his final removal from public life. He retired immediately to Berkeley, and never, except for a day or two, quitted it again. His spirit, wounded and subdued, dwelt with affectionate recollections on the memory of her whom he had lost ; but it is not to be supposed, that he spent his time in unavailing sorrow; for it will soon appear that the latter years of his life, though darkened by domestic affliction, were passed in perfect confor-

mity with that devotion to the pursuit of useful
knowledge which distinguished his early years. The
following strikingly proves the truth of this statement.

Berkeley, Sunday Night.

MY DEAR BARON,

I know no one whom I should like to see here better than
yourself; and as often as you can find a little leisure, pray
come, and exercise your pity. I am, of course, most
wretched when alone; as every surrounding object then the
more forcibly reminds me of my irreparable loss. Every
tree, shrub, flower, seems to speak. But yet no place on
earth would at present suit me but this, and I trust my
friends will not endeavour to take me away; for, strange and
contradictory as it may seem, the bitter cup has a kind of
relish in it here, which it could afford no where else.

Give me a task, and I will execute it as well as I can. Tell
me which subject you want first. Put it down on a slip of
paper when you come. I mean a list of what I promised
you.

God bless you,

Sincerely yours,

E. JENNER.

One of the first events which called him from his seclusion was of a very tragical description. A desperate
conflict had taken place between a gang of poachers
and the gamekeepers * of Lord Segrave, in which one
of the latter lost his life. Dr. Jenner, as a magistrate,
was obliged to exert himself on this occasion, and he
assisted materially in procuring and arranging that
chain of evidence by which the guilt was most clearly

* In January, 1816.

brought home to the murderers; two of whom suffered the extreme penalty of the law, and their companions, twelve, I believe, in number, were transported for life.

Lord Segrave distinguished himself not less by his personal intrepidity than by his judgment and discernment in detecting and capturing the leaders of the gang.

Dr. Jenner attended the trial at Gloucester, and honoured me by staying in my house. He was very much affected by the result of it. Most of the culprits were young men, and the sons of respectable farmers; and though he laboured for the punishment of the guilty, he could not but lament the consequences of the tragedy, which carried such lamentation and woe into so many families in his neighbourhood.

During one of my visits to him at Berkeley in the month of October of this year, I was seized with violent rigors, head-ache, and all the signs denoting the approach of a severe and acute disease. I made preparations for my speedy return home, but luckily for me I was prevented, and to his determination on this occasion, and his subsequent kind and judicious medical treatment, I probably owe my existence.

The disease turned out to be inflammation about the pharynx, the fauces and the tonsils. During the whole of it I had many, many opportunities of witnessing the admirable qualities of this truly great man. His assiduities to myself were unceasing. He punctured my throat three different times; and as an ordinary lancet was rather too short for the pur-

pose, devised an ingenious contrivance for obviating
this difficulty.

While I was at his house an express came to him
from Bath, announcing the alarming illness of his
friend Dr. Parry. He went off the next morning and
returned the same evening, as he was uneasy about
me; I having been delirious in the night.

Dr. Parry had been seized with apoplexy. He
knew his friend Jenner when he went into the room.
" He looked at me," said he, " earnestly for some
time, then grasped my hand, and by piteous moans
and signs expressed how strongly he felt his situation."

In the course of the following summer Jenner had
a considerable illness. I found him very much pulled
down by it. He had neglected himself. He was
walking in the garden when I arrived.

We had a great deal of conversation respecting vaccination and the conduct of the National Vaccine Institution ; the Board having refused to attend to his
cautions touching the interference of cutaneous
diseases with the progress of the vaccine vesicle.
" I am afraid," he observed, " that the extreme
ignorance of medical men on this subject will
destroy the advantages which the world ought to
derive from the practice." I confirmed this, by
what I had lately seen. " I want," he continued, " my
medical friends to rally round me, and to propose
some scheme for the more effectual diffusion of the
genuine information that is required. The present
constitution of the National Vaccine Institution is
bad. The Marquis of Lansdowne and myself had
arranged an excellent plan ; but the change of the

ministry knocked it on the head, and George Rose and Sir Lucas Pepys concocted the present imperfect scheme." Altogether he seemed rather dispirited, and somewhat disquieted by reports of failures, and by the disingenuous conduct of many. I slept at Berkeley and saw him in the morning. He was in bed, and I strictly enjoined him not to go out as he had done. "Tell Catherine and Robert so," said he, "for they think I am shamming; and they would drag me about to the Ridge, and I don't know how many other places." I soon had cheering accounts, as the sub-joined letter will show.

MY DEAR DOCTOR,

Having just heard that a person is going from hence to Gloucester to-morrow morning, I write to tell you that our "stern alarms" will soon be " changed to merry meetings."

I am in every respect better. The pain in my head gone; my respiration easy; expectoration lessened; but there is now no impediment to the separation of the mucus from the membranes. Still, however, there is that susceptibility that a single inhalation of air colder than the temperature of my bedchamber instantly makes me hoarse. I took several doses of the squills, but have omitted it from its singular effects. It called to my remembrance the terrible conse-quences of the dose of cayenne, by giving a glow that almost called up pyrosis. The difference, however, was as great as one to fifty. But does not this indicate a state of stomach that calls for some repairs? I mentioned to you that eating or drinking any thing hastily produced palpitation, and that this has been the case with me for some time past. This palpitation entirely ceases when digestion has gone on to a certain extent. Flatus in the stomach has the same effect. It first came on some years ago from a fright, but I think I have felt more of it within these last twelve months than usual: much of it depends on the state of mind. Depression is sure to produce it, and spontaneous exhilaration (not wine) to take it off. We must talk a little of this on a future day. I did not leave my bed till yesterday afternoon, and had no conception I should have suffered such a diminution of muscular strength. To-day I feel a great increase of strength, but was excessively faint till I took a little animal food, and a small quantity of wine, largely diluted with water.

I am thinking of taking some infusion of colomba with soda, as a tonic suitable for such a stomach as mine. But, perhaps, I shall see you ere long. I am tired. So adieu, my dear Baron, with best affections, truly yours,

E. JENNER.

9 o'clock, Friday night, 2nd August, 1817.

In the former volume I placed the evidence respecting the existence of the variolæ among the inferior animals, I trust, in its true light. That the disease is sometimes met with in the horse was demonstrated by the observations of Jenner, and corroborated by the experience of Drs. Loy, Sacco, and others. The attentive reader will likewise have observed, that the mistake in considering that disease, which is vulgarly called the grease, as the source of the cow-pox, was subsequently corrected by Dr. Jenner himself. It was shewn that the horse* is liable, as well as the cow, to an eruptive disease of a variolous character; and that that disease, when communicated to man, is capable of affording protection against small-pox, even though it had never passed through the cow. For the most part, how-

* See vol. i. p. 242.

ever, the equine affection was seldom recognised in the dairies except in connexion with a similar disorder in the cows. The last time, I believe, that Dr. Jenner had an opportunity of tracing this connexion was in 1817, and I copy the following memorandum from a manuscript written on the 1st of April of that year.

" Rise and progress of the equine matter from the farm of Allen at Wansell. From a horse to Allen; from Allen to two or three of his milch cows; from the cows to James Cole, a young man who milked at the farm; from James Cole to John Powell by inoculation from a vesicle on the hand of Cole; and to Anne Powell, an infant; from Powell to Samuel Rudder; from Rudder to Sophia Orpin, and to Henry Martin; from H. Martin to Elizabeth Martin. All this went on with perfect regularity for eight months, when it became intermixed with other matter, so that no journal was kept afterwards. Proof was obtained of the patients being duly protected."

I find other entries to the same effect. One on the 17th of May runs thus: "Took matter from Jane King (equine direct), for the National Vaccine Establishment. The pustules beautifully correct." The matter from this source was, I believe, very extensively diffused. I received supplies of it; and it was likewise sent to Scotland. I may mention, at the same time, that some years before this period Mr. Melon of Lichfield had found the equine virus in his neighbourhood. He sent a portion to Dr. Jenner; and I believe it proved efficacious.

In the following year I sent him some equine matter, which I obtained from the hands of a boy who had been infected directly from the horse. In this case the disease assumed a pustular form, and extended over both arms.

April 25th, 1818.

Yesterday H. Shrapnell brought me the equine virus and your drawing, which conveys so good an idea of the disease, that no one who has seen it can doubt that the vesicles contain the true and genuine life-preserving fluid. I have inserted some of it into a child's arm; but I shall be vexed if you and some of your young men at the Infirmary have not done the same with the fluid fresh from the hand. On Wednesday, at half-past five, I am threatened with a batch of fossil-hunters. Halifax heads the gang. Will you trust yourself among such folk? With best affections, yours, my dear doctor, very truly, EDWARD JENNER.

After recounting the history of the variolæ vaccinæ in the former volume, I observe (p. 240), " from these facts, it may fairly be inferred that the disease will hereafter be found among cows in other parts of the world." This prediction has been verified in a very remarkable manner by occurrences in the northern parts of Bengal. In reading the account of this epizootic, as drawn up by Mr. M'Pherson, superintendant of vaccination at Moorshedabad, one would almost think that he was copying the language of Lancisi, or Lanzoni, or Layard, who had witnessed similar pestilences in Europe; and it is not less remarkable that the same names are given to it by the natives, by which they designate the variolæ in the human subject. While I cannot help

rejoicing in this additional testimony to the truth of
the doctrines which I have advocated, I cannot refrain from expressing my surprise that the ample
historical records which completely prove the frequent existence of variolæ among cattle should have
been so entirely forgotten. Mr. M'Pherson's account is altogether so curious and important that I
deem it necessary to subjoin it.

On inquiring amongst the natives, I learned that the cows
in Bengal are subject to a disease, which usually makes its
appearance about the latter end of August or early in September, to which the same names are given as to variolæ in
the human subject, viz. bussunt, mhata, or gotee; and on
the 24th of August I was informed that several cows belonging to a native at Moidapore were affected. I consequently determined on again attempting to regenerate the
vaccine virus from the original source.

The animals which were at first affected, amounting in
one shed to eighteen or twenty, had been for a day or two
previously dull and stupid; they were afterwards seized with
distressing cough, and much phlegm collected in the mouth
and fauces. The animals had apparently, at this time, no
inclination for food, or, at all events, they were unable to
satisfy their hunger: their sufferings seemed to be greatest
on the 5th and 6th days, when considerable fever and
pustules made their appearance all over the body, especially
on the abdomen, which terminated in ulceration, the hair
falling off wherever a pustule had run its course. The
mouth and fauces appeared to be the principal seat of
the disease, being in some instances one mass of ulceration,
which in all probability extended to the stomach and alimen-
tary canal.

In those cases where the mouth was very much affected,
the animals died apparently from inanition; whereas in

those cases in which the power of mastication, or even of swallowing, was retained, recovery was much more rapid than might have been expected from the previous severe sufferings and reduced state of the animals : the mortality may be calculated at from 15 to 20 per cent.

From the above description of the disease, the board will immediately observe, that it assumes a much more serious complexion in this country than we have been taught to believe it does at home.   I say taught, because, I presume, it has fallen to the lot of few to witness the disease in England; and it must be inferred, from Dr. Jenner's and other medical writings on this subject, that the animal not only continued to secrete milk, but that the milk was used ; while, in this country, the little that is secreted is never made use of, and perhaps, owing to this very circumstance, the guallahs (or milkers) in India are not affected with cow-pox, as is the case with this description of persons in Gloucestershire, and other counties in England, where the disease is most prevalent.

It is an extraordinary fact, and worthy of remark, that, while the cows were thus affected, no case of variolæ amongst the natives in the village presented itself ; and although the people were ardently averse from handling or going much amongst the cattle at the time of disease, still they all scouted the idea of infection, stating they never heard of any one contracting disease from the cow; consequently they were under no alarm on that score.

In consequence of the extreme jealousy with which all my inquiries on this subject were watched by the Hindoos, coupled with my own anxiety to conceal the object in view, I should have found very great difficulty in prosecuting my investigations, had not the disease assumed the character of an epidemic; all the cattle in the neighbourhood becoming affected, and, amongst others, two belonging to one of my own vaccinators.   I had them covered with blankets, leaving merely the udder and teats exposed to the air; on the seventh day, two small pustules made their appearance on the teats of one, which dried up on the 10th, and the crusts were removed on the 12th day.

From these crusts eleven native children were inoculated; no effects whatsoever were produced on six of this number; two had very slight inflammation on the arms, on the third and fifth days; two had considerable local inflammation and slight heat of surface on the fifth, sixth, and seventh days; but no vesicle formed, although there was marked induration round the puncture. The remaining child's arm was slightly inflamed on the fourth morning, and a vesicle was apparent the next day, which continued to increase till the 9th day, when I was much gratified to find that it assumed all the characteristics of true vaccine.

The poor little child, the subject of this experiment, was about five months old, and suffered much from fever for four days, by which he was greatly reduced, but very soon recovered.

Two children were vaccinated from this patient, with the most complete success; but the symptomatic fever was more severe than I have ever observed it in former cases. Five children were vaccinated from those just mentioned, and the result was equally successful; after which no difficulty was experienced in disseminating the disease.

With the view, however, of satisfying myself that the true cow-pox was introduced, I had two of the children who had been vaccinated with the fresh virus inoculated with small-pox, and both were happily found to be secure. Another instance of the preservative powers of the new lymph deserves mention. Five children, in the Gorah Bazaar at Burhampore, were vaccinated, and shortly afterwards were accidentally exposed to variolous contagion, by residing in the same huts where the disease was raging very dreadfully, but not one of those vaccinated was in the slightest degree affected by variolæ.

Many of the children belonging to his majesty's 49th

regiment, and others in the families of the residents, both civil and military, at this station and its vicinity, have been vaccinated with the regenerated virus. My friend, Dr. French (who invariably has recourse to Bryce's test), Mr. Skipton, the superintending surgeon, and several other medical gentlemen have expressed themselves completely satisfied with the result.

It is a gratifying fact, that since the introduction of the new lymph, the symptomatic fever has been more marked, and the natives have much greater confidence in the efficacy of the operation; in proof of which I need merely mention, that the number presented for vaccination, within the last three months, has much exceeded that of any similar period for the previous two years.

Variola has been more or less prevalent in this neighbourhood for the last seven months, and is now committing dreadful ravages in several parts of the city. Many instances are daily presenting themselves of the disease attacking those who have been previously affected, either naturally or by inoculation; and I am credibly informed that several of the latter have fallen victims to this dreadful scourge. It is melancholy to reflect, that a set of ignorant and mercenary beings (such as the Tickadars in this country) are permitted annually to regenerate the disease, and thereby keep up a continual source of contagion, by which thousands of lives are sacrificed.

Accompanying I have the pleasure to send some vaccine crusts and ivory points, armed with virus, taken two days since, from which I entertain no doubt the disease will be readily introduced in Calcutta; and should more be required, it shall be immediately supplied.

I take this opportunity of acquainting the board, that I have applied to government for one month's leave of absence from my station to visit the Presidency, which I hope to reach by the 12th proximo, when it will afford me much pleasure to yield any farther information that may be required on this interesting subject.

Moorshedabad, November 29th, 1832.

(See Transactions of the Medical and Physical Society of Calcutta, vol. vi.)

In connexion with this subject, it is of importance to remark, that the Variolæ Equinæ have recently been observed in Bohemia by M. C. G. Kahlert, M.D. and assistant professor of the veterinary art in the University of Prague. The work containing this memoir was most kindly transmitted to me by Dr. de Carro of Carlsbad. (See Almanach de Carlsbad for 1833.)

In the course of the year 1818 a virulent epidemic small-pox prevailed in many parts of Great Britain as well as on the Continent. At the same time an increased hostility was evinced to the practice of vaccination. Doubts of its efficacy, which had been artfully excited in England, were propagated to other parts of the world. Dr. Jenner received intimations of this kind from a great many quarters, and professional gentlemen of some name took up the opinion of the anti-vaccinists, and almost declared themselves converts to their doctrines. The events which occurred at Edinburgh, and several other places in Scotland, seemed to the persons to whom I allude, to justify this conduct. The small-pox there, was unusually fatal and malignant. It killed a very large proportion of those whom it attacked in the natural way ; and it likewise spread to many who had previously had small-pox, as well as cow-pox. Dr. Jenner believed imperfect vaccination, or some cause which interfered with the regular and complete progress of that affection, to be the main sources of such evils. He admitted that small-pox might succeed perfect vaccination, just as small-pox does succeed small-pox; but the great number of failures which were reported to have occurred, he thought, could only be accounted for by supposing that some circumstances interrupted the proper influence of vaccination on the system. One of these he conceived to be the existence of cutaneous disease; and this led to the publication of his circular letter, which was intended to draw forth the opinions of professional men on this point. There was a very solid foundation for his statements on this subject, though many have hesitated concerning them; but, be this as it may, there can be no doubt of the truth of his main proposition, namely, that where vaccine failures were very frequent, there must have been some imperfection either in the virus or in the progress of the affection. This fact is rendered abundantly manifest by the different degrees of success which have attended the practice of different individuals. After very minute inquiry I do not know of more than six or eight cases of small-pox after cow-pox among all Dr. Jenner's patients. This proportion is probably no more than might have occurred, had he inoculated for small-pox instead of cow-pox.

Some of the difficulties which perplexed the subject in Edinburgh, arose from a degree of uncertainty that prevailed regarding the character of the epi-

demic at its first appearance. Dr. Hennen was inclined to think that the disease was varicella or
chicken-pox, of a malignant character. He soon was
obliged to abandon this notion. It however tended,
I conceive, to render the investigations, which were
carried on at Edinburgh, somewhat more intricate
than they otherwise would have been. Dr. Thompson, indeed, went further than Dr. Hennen; and in
a very elaborate work endeavoured to prove that the
varicella, instead of being a distinct and peculiar disease, as had been generally supposed, was only a variety
or modification of small-pox. He was led to this conclusion from the great prevalence of the eruptive disease among those who had previously had small-pox as
well as cow-pox; and hence he believed it possible,
that in many previous epidemics, where similar events
had taken place, the truth had been obscured by
giving the disease a specific name, such as swine-
pock, or chicken-pock; when, in fact, it was a disease
of a variolous nature and origin;* and like to that
which then raged in Edinburgh, and other places.
This latter position, I think, has been almost demonstrated; and it can scarcely now be doubted that the

---

* Dr. Sacco, of Milan, has taken another view of this question.
See his paper in the Appendix, No. IV. This document was
kindly extracted and translated for me from Hufeland's Journal, by Dr. Prichard of Bristol. Dr. Sacco unquestionably is
the most extensive vaccinator in the world. All that comes
from him on this subject is peculiarly worthy of attention; and
I have much satisfaction in availing myself of his support to
the opinion, which I have uniformly avowed, that the protection afforded by the genuine Variolæ Vaccinæ is not of a partial
or evanescent character.

unwillingness to believe that small-pox could occur twice in the same person, had suggested the expedient of getting rid of the difficulty by giving the disorder a new appellation. Whether the other part of the position has been equally well made out, I shall not presume to decide. My own impression, however, from all that I have seen, is, that the varicella, strictly so called, is a disease *sui generis;* and that it is not of the same nature as variola.

A remark of Dr. Hennen, in his very excellent paper on eruptive diseases, satisfies me, that he was very near the truth when he asserted the affinity between small-pox and cow-pox. It is very singular that medical men should have been so averse to admit this doctrine, which had been so clearly announced by Dr. Jenner; and should have exerted their ingenuity in detecting differences, instead of tracing analogies, which, when duly understood, tend to remove all difficulties.

" So perfectly convinced am I," observed Dr. Hennen, " of the preventing and modifying powers of the vaccine inoculation, that I should never hesitate about employing it, even though it were probable that my patient had imbibed the small-pox infection; nor should I be deterred from the practice by the idle supposition of the nurse that I was too late, or the learned objection of the doctor that the two diseases could not co-exist; *experience very clearly demonstrating, that there is still something in the mutual relations of these diseases to each other that has not yet been satisfactorily elucidated.*"

I cannot of course quote this latter sentiment

without deriving satisfaction from the conviction that so acute an observer had taken that view of the subject which it has been my great object to explain and confirm, both by past history and recent experience. I cannot doubt that had the investigation been pursued by Dr. Hennen, and had all the evidence which has since been acquired been laid before him, I should have found him a firm supporter of the doctrine which I have espoused.

In the former volume I quoted the emphatic language of Dr. Thompson expressive of his opinions of vaccination, after it had passed through the fiery ordeal at Edinburgh. I cannot do less than give some of the sentiments of Dr. Hennen on the same occasion.—" After the most mature deliberation, I must explicitly avow, that nothing has occurred in these cases which has in the smallest degree shaken my opinion of the great and pre-eminent importance of the practice of vaccination ; whether we view it as a preventive of small-pox in a vast majority of cases, or as a most effectual neutralizer of its malignity in the comparatively few instances in which, from some peculiarity of constitution, or some anomaly in the process hitherto not fully developed, it has failed to afford this permanent security.

" On the contrary, it appears to me that the whole series of cases which I have given, present the most triumphant evidence in favour of vaccination, and place in a most conspicuous point of view the infinite advantages to be derived from the process when duly conducted."*

* See Edin. Medical and Surgical Journal, vol. xiv. page 456.

There can be no doubt that during the first years
of vaccine inoculation there had been great carelessness and inattention in conducting the practice. At
this time, too, (1818) there were numerous, and, I
believe, well-founded complaints of the bad quality
of the vaccine lymph itself. Dr. Jenner received
hints of this kind from Italy, from America, and
many parts of England. These unfavourable accounts were generally coupled with some sinister
and injurious rumours touching his own confidence in vaccination. Here, then, we have an accu-
mulation of evils and misrepresentations, which
could not fail materially to annoy him. There was
great clamour about the prevalence of small-pox
after vaccination, a general complaint of the dete-
rioration of the vaccine lymph, and lastly, a repeti-
tion of the absurd statement of his distrust of his
own solemn asseverations. Strange enough! at this
very time, when people were willing to under-
value and undermine vaccination in this country, a
neighbouring nation, somewhat jealous of the fame
of England, put forth claims to that invention. The
manner in which the French managed this affair
has been already stated. I recur to it here chiefly
to mark the peculiarity of Jenner's situation. His
countrymen were depreciating his discovery at the
moment that another people esteemed it so highly,
as almost to make it a subject of national contention.

The greatly exaggerated statements on the subject
of the vaccine failures, and the hesitating manner in
which respectable individuals spoke on the subject,

threatened to lead to a considerable abandonment of the practice. Under such circumstances, the excellent and honourable Sir Gilbert Blane, whose services at the commencement of vaccination have already been commemorated, proved by the most conclusive reasoning, and an appeal to the most authentic documents, that the importance of the vaccine discovery was in no essential points lowered by the failures which were alleged to have taken place.

In order to bring this matter to the test of calculation, he selected four periods, each of fifteen years, for the purpose of exhibiting the comparative mortality of small-pox. The last series comprehended the time in which the vaccine inoculation had been so far diffused as to produce a notable effect on the deaths from this disease. The result of the whole was, that even under the very imperfect practice of vaccination which had taken place in the metropolis, 23,134 lives had been saved in the fifteen years alluded to; that is, from 1804 to 1818 inclusive. "It will be seen by an inspection of the tables," Sir Gilbert adds, " that in that time there have been great fluctuations in the number of deaths. This has been owing partly to the small-pox inoculation of out-patients having by an unaccountable infatuation been kept up at the small-pox hospital for several years after the virtue of vaccination had been fully confirmed. The greater number of deaths in 1805 may chiefly be referred to this cause. Since the suppression * of

* Small-pox inoculation of out-patients was discontinued, April 28, 1808. The small-pox inoculation of in-patients was persisted in till the 30th of June, 1822 ! ! !

this practice, the adoption of vaccination, though
in a degree so incomplete, in consequence of public
prejudice created entirely by mischievous publications, has been unable to prevent a considerable
though fluctuating mortality from small-pox. The
late mortality from small-pox, though little more
than half of what it was in former times, might have
been entirely saved, if vaccination had been carried
to the same extent as in many cities and in whole
districts on the Continent of Europe, in Peru and
Ceylon."

Sir Gilbert next showed, that making every allowance for the adverse circumstances which had been
reported, vaccination, if duly and vigorously practised, was able to control, and even to extirpate,
small-pox. He fully admitted the occurrence of
those cases of mitigated small-pox after vaccination
which were then prevalent. He showed, however,
from their general mildness, that they could scarce
be called failures; for though vaccination sometimes fails in preventing small-pox, it seldom fails to
prevent death.

I should have great satisfaction in entering more
at large into the reasoning of Sir Gilbert, had it been
compatible with the design of this work. This
essay produced a very beneficial effect on the public
mind. It contained a clear, temperate, and authentic
statement of facts; and although it conceded more
to the anti-vaccinists than was necessary, it proved
that even if every case of vaccination were to be
followed by the mild or mitigated small-pox, it
would be an unspeakable benefit to mankind, by dis-

arming that disease of all its virulence and all its
danger. The whole character of this production,
and the just and animated style in which the learned
and venerable baronet advocated the cause of truth
and humanity, was very acceptable to Dr. Jenner.
Although he was not disposed to go so far as Sir
Gilbert, he nevertheless looked upon the production
as so judicious, and so full of sound and incontrovertible argument, that he was very anxious to extend its circulation beyond the limits of the book in
which it was originally printed.*

"From some unaccountable delay," he observes,
"the last volume of the Transactions of the Medico-Chirurgical Society did not reach me till a
few days ago. I dashed at your paper the moment I opened it; and 1 should set no value on
my feelings if I could not with truth assure you,
that its perusal afforded me the highest gratification. It is exactly the thing the public have
long wanted. A statement so clear and so decisive
cannot fail to make a beneficial impression even in
its present state of confinement; but if I may be
allowed to burst the *blue* walls of its prison-house, I
would, with yours and the consent of the Society, set
it free, and give it the liberty of ranging the world
over. Two or three hundred copies might be dispersed
with certainty of great advantage. In such a case
would you like to make any additions? I know you
would not wish to crowd it with examples of the extinction of small-pox, where vaccination had been
universally adopted; but to the powerful ones you

* Medico-Chirurgical Transactions.

have adduced, there is one observation more that
might be made with the most striking effect; and
that is, the absence of the disease from our armies
during nearly the whole of the late war, which took
place in consequence of a mandate from the commander-in-chief, namely, that every recruit on join-
ing his regiment should be immediately vaccinated,
unless he bore the incontestible proofs of his previous security. This, when contrasted with the recollection of the incessant losses by small-pox among
the troops in former wars, becomes a most interesting
fact.

" With regard to the mitigated disease which
sometimes follows vaccination, I can positively say,
and shall be borne out in my assertion by those who
are in future days to follow me, that it is the offspring entirely of incaution in those who conduct
the vaccine process. On what does the inexpli-
cable change which guards the constitution from
the fang of the small-pox depend? On nothing but
a correct state of the pustules on the arm excited by
the insertion of the virus; and why are these pustules sometimes incorrect, losing their characteristic
shape, and performing their office partially? But
having gone pretty far on this subject in my former
letter, I shall not trouble you with a twice-told tale."

This letter is not dated, but I presume it must
have been written about the beginning of August
1819.

CHAPTER VIII.

SUMMARY OF FACTS RELATING TO VACCINATION—ITS INFLU-
ENCE ON THE MORTALITY OF INFANTS, AND ON POPULATION
—PUBLICATION OF THE CIRCULAR—LETTER TO DR. CHARLES
PARRY—PAPER ON THE MIGRATION OF BIRDS.

IT is now nearly forty years since the practice of
vaccination was publicly adopted in England. In
drawing this subject to a close, it becomes me to
revert to the opinions of Jenner, to compare his
doctrines with the experience acquired during the
whole of that long period; and to ascertain how far
they have, thereby, been confirmed or refuted. I
could wish at the same time to devote some attention to questions of a subsidiary nature; to point
out the connexion of his discovery with the increase
of population, and to explain how it may have contributed to an increase or decrease of the mortality
attendant on other affections.

The discussions which have been published in the
first volume respecting the history of the Variolæ
and the Variolæ Vaccinæ, and the illustrations
thereby given of the sentiments and statements of

Jenner, will render the first of these objects of comparatively easy attainment. We are not now perplexed with difficulties respecting the origin or nature of cow-pox; nor is there any room for much speculation as to its prophylactic powers, that is to say, these powers are alike in nature with those belonging to small-pox itself; and, if human science could enable us to detect those changes in the constitution which, for the most part, render an individual who has had small-pox, insusceptible of future attacks, then we might explain why the Variolæ Vaccinæ exerts a similar influence. Such being the fact, the main questions which it behoves us to put are these : 1st, Has the cow-pox retained those distinctive marks which characterised it when it was first discovered? 2ndly, Has the true virus, after passing through the constitutions of hundreds or thousands of individuals, lost in any degree its prophylactic powers? Does, in short, the affection, at this day, wear the same aspect, and produce the same effect, as on its first introduction?

I have endeavoured to calculate the proportions of failures in vaccination ; but there is much greater difficulty in arriving at the exact truth in this matter than might at first be conceived. This difficulty arises more from the imperfection of our observations than from the nature of the subject itself. In situations where vaccination has been performed with great care, the number of cases of failure is extremely small: whereas, the reports of failures from other quarters are so numerous, that, were they not counteracted by better testimony, men's confidence in vaccination might be shaken. It is quite fair to conclude, when great discrepancies occur, that there must be some fault either in the observer or the reporter ; for it is not possible that such differences should exist if the practice had been conducted with equal attention.

Without, therefore, dwelling farther on this subject at present, I believe that the experience of every well-conducted public institution, as well as of every individual who has fairly investigated the present and past state of vaccine inoculation, will justify me in affirming that cow-pox is now what it was at the beginning. There are instances where the same virus has been passed from one human subject to another for more than thirty years, and its transmission during that period must have been through fifteen or sixteen hundred different individuals, yet no degeneration of its properties has taken place. Though the vaccinations which have been performed by Dr. Jenner and his followers in this district fully warrant this statement, it is nevertheless always proper to employ recent lymph from the cow when it can be procured.

I need scarcely observe that results of this kind can have been obtained only where the greatest attention was paid to the perfectness of the virus in the first instance, as well as during every subsequent inoculation. I have already shewn how much Dr. Jenner insisted upon the necessity of this caution, because he knew that the virus might be deteriorated in many ways ; and that it was possible by inoculating with such to produce a spurious or imperfect vesicle which gave no real protection. It is undeniable that a very large number of reported failures are to be ascribed to the ignorance of, or want of attention to, these important points. It must, nevertheless be admitted, that small-pox has occurred after the most perfect vaccination. The number of such cases, as far as the experience of this district goes, I should say, is not greater than that of small-pox after small-pox.* This statement appears to me to be made out from two sources. There are parishes

* There are no sufficient data to enable us to determine the exact proportion of persons that may be attacked with small-pox after having been vaccinated. By far the most valuable document that has appeared in this country has been obtained from the Royal Military Asylum at Chelsea. The return embraces a period from 1803 to 1833. The number of children reputed to have had small-pox previous to admission was 2532 : of which number, 1887 were boys, 645 girls. The number of boys reputed to have been vaccinated previous to admission was 2498; of girls, 562. Making a general total of 3060.

The number who had small-pox, after reputed small-pox, was 26 : 15 boys and 11 girls. The number who had small-pox after reputed vaccination was 24 : 19 boys and 5 girls.

The number vaccinated at the Asylum subsequent to admission was 628 : 460 boys, 168 girls. Of this whole number, only 2 boys and 1 girl caught the small-pox.

The number who died of small-pox at the Asylum was 4 boys and 1 girl ; of these 5 children, 3 had the disease after reputed small-pox, and 2 had never been vaccinated or undergone small-pox before. (See Appendix to Report from Select Committee on the Vaccine Board).

This document, so far as it goes, fully supports Dr. Jenner's assertion, that the protection afforded by vaccination duly performed is quite equal to that afforded by small-pox itself.

in this county where vaccination has been assiduously and skilfully performed from the commencement. They have enjoyed an almost complete
immunity from small-pox, even though this disease
has been raging in the surrounding neighbourhood.
Again, it has been found that, when small-pox has
made its appearance, and spread epidemically, it
attacked both those who had had small-pox previously, and those who had been vaccinated. This
single fact, if there were none other, proves the
affinity of the two affections ; and the identity of the
laws by which they are governed.

The records from which a body of medical statistics may be collected are yet far from being complete.
Facts enough, however, have been accumulated to
enable us to prove that the mean duration of human
life has very much increased in those countries
where civilisation and medical science prevail.

From accurate calculations made in this country,
it would seem that there has been a surprising increase in the duration of life within the last fifty years.
In 1780 the annual mortality of England and Wales
was calculated to be *one* in *forty;* in 1821, the
yearly mortality was *one* in *fifty-eight,* or *one* in
sixty; so that in *forty* years the mortality has
decreased nearly *one-third.* In order to point out
the connexion between this decrease and the practice of vaccination, it is proper to observe, that the
ratio has materially increased since the year 1801.
In that year the calculation was *one* in *forty-seven,*
affording only a difference of seven between that and

the preceding twenty years : whereas, the difference in the succeeding twenty, that is to say, from 1801 to 1821, was no less than thirteen.

It is not easy to ascertain precisely how much of this addition to the duration of human life is to be ascribed to the influence of vaccination. The great improvements that have taken place in all the physical sciences, the increased attention which has been bestowed on Medical Police, the advances that have been made both in the knowledge and practice of medicine itself; all have contributed to these interesting results : but the effect of controlling or subduing the mortality of such a scourge as smallpox, must have been great and extensive. Leaving this part of the subject for the present, I shall proceed to elucidate another, for which more precise and accurate data have been afforded. The mortality of children has always been a painful subject of meditation. This mortality varies considerably in cities and towns, villages and country parishes. Some very valuable information has been published on this topic by Mr. Roberton, of Manchester, in his work on the mortality and physical management of children. According to his tables it would appear that the mortality of children under ten years of age, in cities and large towns, was 51.39 ; in smaller towns and cities, it is 48.97 ; in village parishes, 49.90 ; in agricultural parishes, it is 35.40. It would appear that the mortality of those under *ten* in foreign countries approaches that of our own. In the German tables, it is reported to be 43 per cent. for the country, 47.7 for small towns, and 50.2 for large cities. Duvillard gives 44.89 as the average of the kingdom of France, which is within a fraction of the rate of infantile mortality in this country, it being 44.91.

It is somewhat remarkable that the same ratio of mortality in after-life does not hold between this country and those just mentioned. If the calculations made, at least, be correct, it would seem that while the average annual mortality in England and Wales is about *one* in *sixty*, that of France is *one* in *forty*. In Prussia and Naples it is stated to vary from 1 in 33 to 1 in 35. In Wirtemberg it is 1 in 33. The countries which are said to approach nearest to England are the Pays de Vaud, Sweden, and Holland: the first affording an average of 1 in 49; the latter two, 1 in 48. Should the facts really be as above stated, they afford astonishing testimony to the superior salubrity of England over the most favoured spots of Europe.

Notwithstanding the proofs of the power of vaccination in diminishing the mortality from small-pox, it has been a question whether *infantile* mortality has been diminished; it having been supposed that the beneficial effects of vaccination were countervailed by a greater mortality in the other diseases of children. This very discouraging statement was published by Dr. Watt, of Glasgow, in an appendix to his work on Chincough. This opinion, which was hastily adopted and unwisely promulgated, has unquestionably had a great effect in retarding the progress of vaccination. It, unfortunately, gave countenance to some of the worst prejudices of

those who were opposed to the practice. It has
now, however, been demonstrated that his conclusions rested upon a fallacy which ought to have been
avoided. He expressed himself as utterly astonished
to find the number of deaths under ten years as
great in 1812 as it had been in 1783. In making
this calculation he seems to have forgotten, that
Glasgow, during that period, had more than doubled
its working population. This important fact is very
fully illustrated by Mr. Roberton. In the year
1783, the per centage of deaths under ten years for
Glasgow, was 53.48, while the annual mortality of all
ages was 26.7. In the six years preceding 1812,
the per centage of deaths under ten years is 55.49,
and the annual mortality of all ages is 1 in 40.8. It
thus appears that the relative proportion of deaths
under the tenth year is greater in the latter than in
the former period : but, if we take into account
the increased population of that city, it will be
seen that the actual mortality under ten years, in
1812, was nearly one-third less than in the first-
mentioned period. It is somewhat remarkable, that
the tables published by Dr. Watt himself did not
lead him to the correction of his error. In his first
period, namely, from 1780 to 1785, the population of
Glasgow was reckoned at 44,360, and the mortality
was 1 in 26.7 ; while in the last-mentioned period,
namely, from 1801 to 1811, the population had
amounted to 96,977, while the mortality was only
1 in 40.8.

Dr. Casper, in his essay on the mortality among
children in Berlin, has fully proved, that whilst the

mortality from small-pox has been evidently diminished, that from other diseases has, in like manner,
lessened. In twenty years antecedent to the first
introduction of cow-pox, the deaths under puberty
were as 51 to 100. In eight years succeeding 1814,
the deaths were only 42 in 100. Dr. Casper was
not satisfied with these general results; for he has
proved, by the most direct and conclusive evidence,
that the other diseases of children have really become less fatal as vaccination has been more generally employed. In four years previous to 1790, the
deaths among children from all diseases, exclusive
of small-pox, were 39 in 100. In four years previous to 1823, they were only 34 in 100. According
to the same authority, the two diseases, namely,
measles and scarlet-fever, which were supposed to
have increased in severity since the introduction of
vaccination, have been, in fact, neither more prevalent
nor more fatal, but rather an actual diminution of
the mortality from measles has taken place. All that
I have been able to learn in this district fully con-
firms this statement. The healthiness of children
has been augmented rather than diminished; and if
in any epidemic of measles or scarlet-fever a greater
mortality should arise, it may fairly be ascribed
to the increased number exposed to its influence,
rather than to any exacerbation in its type or character.

Since the introduction of vaccination the mortality of those under two years of age has been greatly
diminished; but between this age and that of ten, it
is affirmed that the mortality has proportionally

increased. The inference drawn from this statement has been, that as small-pox in former times
chiefly extended its ravages by sweeping away children under two years, what is gained in the saving
of life by vaccination, is lost by an increased mortality from the other infantile diseases. It seems
that a greater number of children now die of measles
than formerly; but there is no satisfactory proof
that this disease itself has become more severe since
the practice of vaccination. There is, in short, no
evidence that in a given number of cases of measles
the mortality is greater than when small-pox was
prevalent. It is clear, that unless facts of this kind
can be adduced, the argument is of no value; and
the phenomenon can be explained on more accurate
principles. Formerly, a variolous epidemic so completely swept away the population under ten years
of age, as to leave comparatively few to be exposed
to the influence of other epidemics. Mr. Roberton
has confirmed this statement by authentic documents
deduced from the history of variolous epidemics in
Warrington and Chester. In the latter place, Dr.
Haygarth reports that there were, in the year 1774,
546 deaths; of these, 334 were under ten years, and
202 were caused by small-pox.

This part of the subject has been very ably
treated by Mr. Edmonds. From his tables it appears that the mortality in England under five years
of age is now only half as great as it formerly was supposed to have been. He has shewn that this is the
case both *absolutely* and *relatively* to the mortality
at all other ages. It is believed by the best authori-

ties, that before the introduction of vaccination, there died under the age of five years, out of one hundred born, sixty in London, and forty in all England. During the twenty years ending with 1830, there died under the age of five years, out of one hundred born, only thirty in London, and twenty in all England.*

On the whole, I feel perfectly assured that vaccination has not only had a direct and positive influence in subduing or diminishing the mortality from small-pox, but that it has, likewise, had a beneficial effect in maintaining the human constitution against the attacks of other diseases. There is much reason to believe that small-pox left those whom it attacked, much more susceptible of illness. Scrofula, for example, in all its forms, was certainly very often excited; and, in particular, pulmonary consumption. The time, perhaps, is not yet arrived for drawing accurate conclusions regarding the increase or decrease of such diseases. It, nevertheless, appears from the London Bills of Mortality, that since the year 1808, the deaths from pulmonary consumption have been decreasing, and it is undeniable, that the mortality from all diseases above ten years, is nearly as much diminished as that under ten.

I might fill a volume in recording the evidence that vaccination, when extensively and efficiently performed, can extirpate small-pox. Wherever the practice has been judiciously followed up, its success,

* See British Medical Almanack, 1837.

in this respect, has been nearly complete. The measures taken by the government of Sweden have already been mentioned in the former volume of this work. The following document will tell with what success.

In the year 1779 the small-pox destroyed 15,000 persons.

1784	.	.	. 12,000
1800	.	.	. 12,000
1801	.	.	. 6,000
1822	.	.	. 11
1823	.	.	. 37

For a period of eight years not a single case of small-pox occurred in the dominions of His Danish Majesty. The whole inhabitants had been vaccinated.

Between 1752 and 1762, the small-pox carried off in Copenhagen alone 2644 victims; from 1762 to 1772, it carried off 2116; from 1772 to 1782, 2233; from 1782 to 1792, 2785; but from the introduction of vaccination, in 1802, to the end of 1818, only 153 persons have died of the small-pox; namely,

1802	.	73	1811	.	0
1803	.	5	1812	.	0
1804	.	13	1813	.	0
1805	.	5	1814	.	0
1806	.	5	1815	.	0
1807	.	2	1816	.	0
1808	.	46	1817	.	0
1809	.	5	1818	.	0
1810	.	4			

158

(See Annals of Philosophy, August, 1819.)

I am not ignorant of the history of the variolous epidemics which of late years have prevailed in different parts of Europe. I know that in Lombardy, in Denmark, and in France, as well as in England and other places, disasters of this kind have taken place.

It has been imagined, from occurrences which arose during these epidemics, that the protecting power of vaccination is weakened by the lapse of years: not having myself witnessed any facts of this description, I cannot decide whether or not the opinion is well founded. The experience of this district, the birthplace of vaccination, does not countenance the idea. All those that I have heard of who were vaccinated thirty or more years ago, appear to have resisted small-pox contagion as much as if they had previously had that disease. I may also add, that I have never seen but one fatal case of small-pox after vaccination, during the whole of my professional life, although my acquaintance with the medical gentlemen of this and the adjoining counties was very extensive, especially when I held the appointment of Physician to the General Infirmary. As far, therefore, as my knowledge goes, I would repeat Dr. Jenner's maxim, and say, that vaccination, duly performed, will protect the constitution as much as small-pox itself.

It can scarcely, in my opinion, be doubted, where small-pox has *prevailed* after cow-pox, that there has been some imperfection in the vaccine process, and that thereby another maxim of the author of vaccination is illustrated, which tells us, that these imperfections may be propagated, and that they will afford varying degrees of protection, according as they recede from, or approach to, the perfect standard.

I am especially struck with the force of these remarks when I look at some of the recent accounts from Denmark. From Dr. Wendt's book, it appears that re-vaccination is employed in the Danish army to counteract the contagion of variola. He mentions, that in the year 1835, out of 3173 persons between the ages of twenty and twenty-five, 2175 were successfully re-vaccinated, while 998 resisted the infection. This is an enormous proportion ; for every one of the larger number would have been liable to an attack of small-pox. I cannot avoid thinking, therefore, that the first vaccinations had been imperfect.

The results of re-vaccination in the Prussian army have likewise led to the belief that the succeptibility to small-pox among the vaccinated is annually increasing. Thus, of 100 re-vaccinations in 1833, 31 were successful ; 37 in 1834 ; 39 in 1835; and 43 in 1836. How it may be in other parts of the world, I cannot say ; but I am constrained to repeat, that nothing has happened in this vicinity to countenance such a statement. I cannot, therefore, arrive at any other conclusion than that the defect is not in vaccination itself, but in the manner of conducting the process, or in the employment of imperfect virus.

Amid all these discouraging circumstances, it is with the greatest satisfaction I refer to the bills of mortality for London during the year 1837 ; which give only 217 deaths by small-pox. This is the smallest number that has ever been recorded in any one year since the first establishment of this register. It would be a glorious triumph if, through the exertions of the different vaccine establishments, London should be freed even for a season from this scourge.

It is calculated that within the last fifteen years the population of Europe has been augmented by nearly twenty-nine millions.* The cessation of hostilities and the cultivation of the arts of peace will, in some degree, account for this great increase. It will, nevertheless, be found that the removal of one prolific source of mortality, by the introduction of vaccination, has tended materially to this result. In Prussia, where the practice was very early adopted, it is proved by accurate statistical investigations, that from the years 1817 to 1827, the increase of population amounted to 1,849,561. In this period are included some of those years in which Dr. Casper demonstrated the great diminution in the number of deaths among children at Berlin.

In Sweden and in Denmark, where vaccination was also adopted and enforced by the influence of the government, population has been making rapid strides. In the latter country the increase is calculated at the rate of two per cent. In Sweden it is not quite so much. In European Russia it is supposed that from the year 1815 to the present time 7,000,000 have been added to its population. It is important to remark, that this increase is not so

N.B. This was written in 1831.

much owing to an access of births, as to the diminution in the number of deaths.

In Austria the results are as striking and satisfactory. According to the returns published by the
geographical board at Vienna, it is inferred that the
increase on the population of 1815, has in twelve
years amounted to more than 27 per cent. The
population being in 1815, 27,000,000; more than
7,000,000 have been since added to its numbers.

It is a remarkable fact that the rate of increase
has been slower in France than in many other parts
of Europe. From the work lately published by the
Baron Dupin, it appears, that the annual increase on
each million is 6536. This gives a total annual
augmentation of about 200,000.

The rate of increase in Great Britain and Ireland
has been nearly double that of France: Great Britain having increased by 200,000 annually; and
Ireland in at least the same ratio. So that the
actual increase in the population of the United Kingdom has in the same space of time equalled that of
France.

The following statement, taken from the work of
M. Dupin, affords an interesting view of the rate of
increase in the population of the principal states in
Europe.

Annual increase upon each million of inhabitants.

Prussia	27,027	Russia	10,527
Britain	16,667	Austria	10,114
Netherlands	12,372	France	6,536
Two Sicilies	11,111		

M. Dupin ascribes the low rate of increase in

France, in some degree, to the neglect of vaccination, particularly in the south, which appears to be much behind the north in every improvement. It is, however, most important to repeat that, though the number of inhabitants in France has not increased in the same ratio with other countries, the degree of mortality is materially diminished. The births are said to be less numerous than they were in the year 1780; and yet it is affirmed that the annual addition made to the population is greater by 44,000 than it was at that period.

M. Berard gives the same cheering statements, though somewhat in a different way. In 1780 the deaths annually in France were as 1 in 30. From 1817 to 1824 they were as 1 in 40: while it appeared that, during the same period, the difference between the deaths and the births was nearly 200,000 in favour of the latter. The cause of this beneficial alteration in the rate of mortality is well illustrated in the Memoir of M. Benoiston de Chateauneuf, who has proved that, in 1780, fifty out of every hundred new born infants died in the two first years of life; at present, only 38.3: which gives an augmentation of one-fourth (of lives) in the hundred. It is during this infantile period of life that the influence of vaccination is most sensibly evinced, and much of this favourable result is, doubtless, to be ascribed to it. In the subsequent periods of life, results nearly as gratifying have been ascertained by M. Chateauneuf: thus, formerly, before ten years of age, 55 died; at present, 43.7 : 21.5 reached the age of fifty; now 32.5 : fifteen attained the age of sixty; now, twenty-four.

It is admitted on all hands that our population has been increasing rapidly since the beginning of the present century. There has not only been a much larger proportion of births ; but, what is more to my purpose, a considerable diminution in the proportion of deaths. In the early part of the preceding century the proportion of births to deaths, in three successive decades, was as follows.

Years.	Births.	Deaths.	Proportion.
1710	143,735	145,146	900 to 1000
1720	159,906	165,956	963 to 1000
1730	166,514	182,579	912 to 1000

After the year 1740 this losing account began to alter, and the proportion of births to deaths to increase steadily, but slowly. In the year 1800 the births were 263,408, the deaths were 193,476 ; giving a proportion of births to deaths of 1361 to 1000. In the year 1810 the ratio had increased to 1518 to 1000. And in the year 1820, the numbers stood thus : births, 334,007 ; deaths, 208,153. Proportion, 1605 to 1000.

The same results are still more strikingly exemplified by a reference to the population of England and Wales since the beginning of the last century, as deduced by Mr. Rickman from the births only. In fifty years, namely, from 1700 to 1750, the increase amounted only to 900,092 ; whereas in fifty years, from 1770 to 1821, the increase was no less than 4,657,000. Great as this increase is, it is rendered more remarkable in that by far the greatest amount

s 2

of it has taken place within the last twenty years of the period, nearly three millions having been added in that time.*

It is impossible to say what may have been the exact effects of vaccination in producing these results ; but that they must have been very great no one can doubt who is aware of the comparative mortality of infantile diseases, and the great proportion that it bears to the total number of deaths at all ages.

It is at once my duty, while it yields me pleasure, to present the gratifying details which have just been closed, in connexion with the personal feelings and character of Jenner himself. Unhappily, his race was run before the full tide of gratulation, with which the discoverer of vaccination might now be hailed, had reached his ear. The reader, however, cannot fail to observe that, even from the outset, Jenner s confidence was as firm as it was just and well-founded. But, after all, it is a marvellous subject ; and even those who have watched its progress, who have anticipated and longed for the success that has attended it, cannot, when the accumulated evidence is brought fully to bear upon the lives and happiness of kindreds, and tongues, and nations, avoid wondering at the signal mercies with which, in these days, Providence has crowned the exertions of one of our fellow creatures.

* See evidence on the poor-laws taken before the House of Lords (1831) ; but especially the tables and remarks delivered in by John Barton, esq. See also the excellent work on medical statistics by Dr. Hawkins.

Wonderful as have been the events of the last years, great as has been the advancement of human knowledge, and rapid as has been the progress in all the arts calculated to promote the comfort and convenience of life, the most remarkable phenomenon is, certainly, not that we have subdued the elements to our use—not that we can multiply at will the products of our ingenuity—not that we have brought mechanical agents to take the place of active and intelligent beings; but that we have been enabled to stay the power of death—to keep him for a season from his victims—and to say that the day of grace and preparation has been lengthened.

In thus exulting in the benefits of vaccination, it will be seen that I cannot fall in with the lamentations of the whining economists who look upon an increase of population with an evil eye, and permit selfish and limited views of what is best for the well-being of the community to interfere with the richest blessings of Providence to man*.

As we approach towards the conclusion of Jenner's life, his opinions, whether on vaccination or any other subject, assume a more solemn and impressive character. The matured reflections of a well-disciplined mind, conscientiously and perseveringly directed to any scientific object, will always command attention; but when the immediate consequences of such inves-

* La Place, in a conversation with Sir James Mackintosh, at Madame de Romford's, said, that the vaccine, when it supplants the small-pox, will add three years to the medium duration of human life.—See Life of Mackintosh, vol. ii. page 322.

tigation touch the lives and security of the public, a deeper interest is added to the inquiry. No man more strongly felt the power of such truths than Jenner : in short, the whole of his labours were preeminently distinguished by their intimate alliance with great results bearing immediately on the well-being and happiness of society. Under a keen sense of his responsibility he wrote, and it is this circumstance which gives to all his communications an air of trustworthiness and sincerity that entitles them to the utmost consideration.

The value of his labours, when gauged by this rule, has by no means been appreciated. It has not been duly felt either by his professional brethren or by the public, that there was a moral grandeur associated with his humility and perseverance. The time, I trust, is coming when justice will be rendered to his memory; when it will be acknowledged that those qualities which give the principal excellence to scientific pursuits did, in a peculiar manner, adorn and dignify his mind. Why, it may be asked, should I recur to this train of feeling? Because it is but justice to him, and holds out to medical men an example worthy of all imitation.

In the greater number of discussions that occupy the attention of professional men, such qualities seldom maintain a prominent place. The play of ingenuity, the contention of wit or learning, the subtle exercises of the understanding, may all in turn divide their care, and afford temporary subjects for amusement or instruction. Such have been the foundations of most medical theories. In general their application to practice is too indefinite and uncertain to render the authors of them amenable to any other than a speculative or intellectual tribunal, whose awards do not necessarily infer any moral delinquency. It was not so with regard to vaccination; and doubtless every word written by Jenner on that subject came from him, not as an individual emulous of distinction, but desirous of advancing truth and promoting the essential well-being of his fellow-creatures. The envy and rivalship that he encountered, tended very much to obscure and counteract these excellences. He was always looked upon as a suspected witness, even though his evidence was corroborated by an abundance of impartial and disinterested testimony. At the outset of the investigation such fastidiousness was not unbecoming; but to see it persisted in after every reasonable doubt had been removed, and with no other effect than to damage a good cause, and to inflict unnecessary pain on a just and generous man, is a lamentable proof of the blindness of the understanding, and of the obliquity of the heart.

The feelings of hostility towards him became more virulent and vindictive with every alleged case of failure. He was held answerable for every supposed misadventure of this kind, while the real merits of his discovery were artfully overlooked, and his cautions and admonitions, in too many cases, disregarded. Under the pressure of popular obloquy of this description, he continued firm to his principles, and vindicated his doctrines with mildness and wisdom.

" I have searched" (says he, writing to a correspondent) " in vain for a record respecting the person you name to me, who has had the small-pox after being vaccinated by me ten years ago. From the date you fix, it probably took place at the time I permitted persons of all descriptions, not only those of the town, but from the districts around, to come to me weekly. The small-pox was at their heels, and this drove them to my house in immense numbers. I was literally mobbed, driven to a corner, and made a prisoner, necessitated to submit to their will. ' *The man shall do me next.*' ' *No, he sha'n't ; he shall do me,*' was the language I was often obliged to hear and submit to. For many successive inoculating days the numbers that assembled were, on the average, about three hundred. The taking of notes, or the observance of anything like order and regularity, was out of the question. However, I persevered with patient submission, and completely gained the grand point I aimed at : the small-pox was subdued in every direction. Though this was the fortunate result, yet it would be absurd to suppose that out of this vast body all could go through the disease with that correctness which protects them from small-pox infection ; in numerous instances, indeed, they did not afford me an opportunity of judging of their security by ever returning to shew me their arms ; and this teazing occurrence not unfrequently happened among the common people of Cheltenham, when I vaccinated on a reduced scale. But now, sir, more immediately to the consideration of your communication. Let us admit that the individual in question went through the vaccine in all its stages with the most perfect regularity, and that, at the expiration of ten years, she became infected with the small-pox, and had that disease with as much regularity as if she never had been under the influence of vaccination. What then? Is the small-pox itself a perfect and constant guarantee against future infection? Where is the medical man, possessed of experience in his profession, and of an inquiring mind, who will not answer this question in the negative? Cheltenham is certainly not exempt from this deviation in a general law of the animal economy, as it exhibits abundant testimony of the contrary; one instance, indeed, is so very remarkable, that it is worthy of being recorded: I allude to that of the lady of Mr. Gwinnett, who has had the small-pox *five* times.

"When the small-pox appeared in Cheltenham two summers ago in one of the lanes which leads out of the High Street, though the situation was exposed, and numbers of children who had been vaccinated were within reach of the infection, yet none of them took the disease. However, it happened that a young fellow, who had been inoculated some years ago for the small-pox at Upton, was not so fortunate. He became infected, and had this distemper with some degree of severity. His case, by the way, was one of those which illustrates the truth of my observation respecting the cause that proves

an impediment, in either inoculation, to that con-
stitutional change which nature demands as a safe-
guard against future infection. This young man, at
the time of his being inoculated, had Tinea Capitis,
and for some years after.

" Not long since I was called to a footman in a gentleman's family, whose case was precisely similar to
that just stated, except that it was more severe. On
examining the inoculated arm, I found the cicatrix
more extensive than usual, which convinced me of
some irregularity in the progress of the disease.
And here I will observe, as a good practical remark,
if you are ever called upon to form a prognostic as
to safety in a vaccinated child, that a cicatrix much
beyond the usual boundary, should always be looked
upon with a suspicious eye ; for it is impossible that
the progress of the vesicle can have been correct
when this appearance presents itself; and in the
vaccine, you know, the appearance of the arm is our
only guide.

" I find myself imperceptibly drawn into practical
remarks ; and, as this is the case, I will mention one
more, on which I cannot lay too much stress, namely,
robbing the vesicles too frequently of their contents.
How often have I seen, where there has been but
one only, that this poor solitary thing, which is expected to perform an office of such immense importance, has been cut and mangled day after day in the
rudest manner ! This has not only happened in the
early periods of the vaccine practice, but the evil
still exists, as I witnessed it on an infant in Cheltenham not many months before my departure ; and I

have been doomed again and again, in spite of all my remonstrances, to be a spectator of this dangerous practice both in London and elsewhere."*

The foregoing very interesting letter (written in 1817) contains a summary of all his doctrines, together with some of the most important practical directions for conducting vaccination. The case of the Honourable Mr. Grosvenor has been already particularly detailed. In one of Jenner's note books of this date, I find some additional facts recorded which it is not unsuitable to lay before the reader in this place. " I vaccinated this young gentleman in a puny state of health at about a month old. Lady Grosvenor was timid, and prevailed on me to deviate from my usual mode of practice; and to make one puncture only; and the pustule it excited was unfortunately deranged in its progress by being rubbed by the nurse. The small-pox, which followed, went through its course in a shorter period than usual, and scarcely left any mark." Again, he observes to another correspondent about the same time, " The failures of small-pox inoculation far exceed those of the vaccine in those districts where I have vaccinated on a large scale. This is, in proportion to the numbers, as the latter, I imagine, exceed the former by ten to one; and of what import are the few that have occurred, as they have not produced fatal consequences?" Let us contrast these statements with his remarks on one of those varioloid epidemics, as they were called. " I wish it were in my power to give you any information respecting the circum-

* Letter to Dr. Coley, from Dr. Jenner's Journal, 1815 to 1820.

stance you mention at East Sheen. I have not been
within a hundred miles of the spot these three years.
But it strikes me as very extraordinary that more
than twenty individuals assembled together, and
who came from different districts, and consequently
were vaccinated by different practitioners, should all
have had this eruptive disease, which was called the
small-pox. I really cannot conceive that such a
thing could have happened, if they had been vaccinated by a cobbler. I am told, too, that the usher
of the school shared the same fate as the boys."

The purport of these extracts is to show, first,
Dr. Jenner's great carefulness in conducting the
process of vaccination; and secondly, his consequent
success. Although he could not always watch the
progress, I have reason to know that the estimate
of his success, as given above, has not been falsified
by disastrous events since his death.

I will now submit to the reader an extract from a
letter to Sir Gilbert Blane, written in 1819, which
more immediately bears upon the varioloid disease
as it appeared in Scotland about that time. " I
have often said, and I still declare it, that if ever
anything occurred which militated against my early
assertion respecting vaccination, namely, that if
properly conducted, it would afford a security
against small-pox as perfect as the inoculation of
small-pox itself, I would immediately proclaim it
to the world. The principle of vaccination is good,
it is immutable; but its application has been bad,
and continues to be so. The practice is conducted

heedlessly in many respects, but chiefly with inattention to a subject I brought before the public so long
since as the year 1804. The paper, a copy of which
I now send you, came out that year in the Medical
and Physical Journal for the month of August."

The minuteness of his attention to every deviation from the correct progress of the vaccine vesicle
will clearly explain the reason of the all but uniform success of his practice. I wish to impress
this fact most strongly; because every medical man,
by similar caution, might obtain the same results.
I could add many extracts from his journals and
letters, all proving his unceasing vigilance and
caution. The state of the virus to be inserted, the
condition of the skin of the person about to be
vaccinated, the character of the vesicle itself, and
the necessity of allowing one, at least, to run its
course undisturbed, were points uniformly insisted
on. In compliance with the doctrines often stated
in these volumes, and when alarm possessed some
minds for the security of their offspring, and would
have induced them to put that subject to the test by
the employment of small-pox inoculation, he invariably dissuaded from this practice. He considered it
dangerous to the individual as well as to the community. He was aware that equal security and less
risk was to be obtained by re-vaccination. Out of
numberless proofs of this kind I select the following
from one of his journals written nearly twenty-six
years ago : " Whenever there is a shadow of doubt
upon the mind respecting a child being perfectly

vaccinated, I always recommend the insertion of a little vaccine fluid."*

Another fact illustrates the accuracy of his observation and the correctness of his judgment. His doctrines led him to believe that whatever offered an impediment to vaccination would stand in the way of variolation likewise. " Vaccinated Richard Stephens, a recruit in the Gloucester Militia, in one arm; Henry Jenner, in the other arm, aged 19. He has Strophilus Pilaris of Willan, and has been exposed to small-pox without effect. This man remained uninfected." (Journal, February 1812.)

This was one of the facts which induced Dr. Jenner to believe that cutaneous affections interfered with the progress of vaccination. But why refer to private journals and unpublished documents? The same thing has been announced times out of number by Dr. Jenner in every form. The following sentence concludes a pamphlet first published in 1808, and republished verbatim in 1811. " At the commencement of vaccination, I deemed this test of security (i. e. the insertion of small-pox matter) necessary; but I now feel confident that we have one of equal efficacy, and infinitely less hazardous, in the re-insertion of the vaccine lymph."

When I read these and other most plain and intelligible propositions laid down by such a man as Jenner, and find those who ought to be better acquainted with the subject reasoning as if they had

* Letter to Mrs. Fleet, Darent, near Dartford, Kent, November 17th, 1811.

no existence, disregarding his accumulated experience, and disfiguring a simple and beautiful system by their dogmatical and unsatisfactory commentaries, I almost despair of the successful progress of scientific truth in our profession. But the very same fact was announced at a still earlier period, more hesitatingly, it is true, but still with sufficient clearness and form to mark well the mind of the writer. The very first paper of instructions published, I believe, by Jenner in the commencement of 1799, contained these, among other important statements.

" A little practice in vaccine inoculation attentively conducted, impresses on the mind the perfect character of the vaccine pustule; therefore, when a deviation arises, of whatever kind it may be, common prudence points out the necessity of re-inoculation, first, with vaccine virus of the most active kind, and secondly, should this be ineffectual, with variolous virus. *But if the constitution shows an insusceptibility of one, it commonly does of the other.*"

For some time before his death he was employed in reviewing his own opinions, and in comparing them with the facts which had been obtained from the experience of his brethren throughout the world. Had his life been spared, it was his intention to have presented to the public a digest of the whole, matured, and, as far as possible, perfected by his own uninterrupted investigations. Though this his deliberate judgment has not been thus set down, it is satisfactory to know that enough has been recorded to leave us in full possession of all his views.

The reported failures of vaccination, and the occurrence of several violent variolous epidemics in different parts of the country, induced him to endeavour to rouse the attention of professional men to those points in the practice of vaccination which he deemed essential to its success. With such intentions, he printed a circular letter early in 1821, which was sent to most of the respectable medical men in the kingdom; in it he directed their observation to the three following questions :—First, whether the vaccine vesicle goes through its course with the same regularity when the skin is under the influence of any herpetic or other eruptive disease, as when it is free from such affections; secondly, whether the existence of such eruptive diseases causes any resistance to the due action of vaccine lymph when inserted into the arms ; thirdly, whether cases of small-pox after vaccination had occurred to the observer ; and if so, whether such occurrences could be ascribed to any deviation in the progress of the vaccine pustule, in consequence of the existence of herpetic or other eruptions at the time of vaccination.

From personal knowledge, I have no doubt that his doctrines on this point are, in the main, perfectly correct. There are some who think and teach differently; but to say the least of it, this is neither wise nor prudent. It is admitted on all hands, that the progress of the vaccine pustule, from its first appearance till it has done its office in the constitution, is a delicate and a very important one. That so small an outward appearance should produce such extensive changes in the animal frame, is one of the

most remarkable phenomena in pathology. It is
likewise admitted by those who do not hold Jenner's
opinions, that the more decided interference with
the vaccine vesicle will mar its full salutary effects;
why, therefore, it may be asked, may not a less perceptible disturbance produce similar results? As it
is not my purpose or object here to enter into minute
detail, and as I am much more anxious about sound
and useful practice, I would again strongly urge all
who conduct vaccine inoculation to look upon the
subject in a plain practical point of view. Seeing,
therefore, that our aim is to rescue the constitution
from the attacks of an extremely malignant and fatal
disease, by means of a very slight affection, it is impossible to be too cautious in every thing that regards
the latter.

It is bare justice to Dr. Jenner to exact the performance of these conditions; but it is an act of more
imperative duty as regards the safety of the vaccinated, and the welfare of the community. I would
hope and believe that the effect of the circular has
been to draw men's attention more to these points,
and to prevent that loose, unsatisfactory, and un-
scientific practice, which could not but lead to disappointment, and injure the character of vaccination.

The answers which Dr. Jenner received to his circular were numerous, and in general satisfactory.
I will not refer to any from his professional brethren;
but the following letter from the excellent rector of
Leckhamstead, near Buckingham, speaks so judiciously and wisely, and, besides, contains some
facts so valuable, that I am induced to present it. The

writer, it will be remembered, distinguished himself as an ardent and successful promoter of vaccination ; and his testimony is of great value.

Leckhamstead, near Buckingham, June 29, 1820.

DEAR SIR,

Your letter did not reach Buckingham till June 23rd, though dated the 12th. The object of inquiry appears to be the extent to which cutaneous diseases reject or modify the vaccine virus, so as to render the efficacy and security doubtful. I have looked over a number of copies of communications to Dr. Harvey, and will with great pleasure send you the transcripts of the interference of variolous and vaccine infection, and the superseding power of the latter if applied in time, six of which took place at Old Stratford in 1816, among the children of one family, being the whole time under the same roof. The distress and alarm at that time were extremely great, as the inhabitants were recovering from the measles when the small-pox broke out. The anxiety of the parents was such that I was induced, contrary to my own opinion, to vaccinate several where the fever of measles had not completely subsided: the consequence of which was nothing more than that the vaccine virus lay dormant in its cell till the field was clear, and came into action two or three days later; but afterwards proceeded in as regular and decided a manner as in constitutions which were not previously engaged.

I discovered at a very early period that the itch was not an impediment; as to the shingles, I cannot speak. The grand rejecting agent in infants is the tooth rash, or, as it is here commonly called, the red gum, especially while it continues bright and active. Dr. William Cleaver (when Bishop of Chester) promoted an extensive variolous inoculation in his diocese. Some years after, he asked me if I could account for the very frequent failure of communicating the infection to young children. I told him that it applied equally to the vaccine; though frequently, if the virus was fresh and active, it would be suspended in its career for a time only, but push forward with success at last.

I beg to assure you, Sir, that nothing I have met with has, in the slightest degree, shaken my faith in the vaccine. I have seven children, the eldest sixteen, all vaccinated by myself; and of 14,305, all within a few miles of this place, I have never heard of a single fatal disappointment; and of only two or three cases of modified, or what I should feel inclined to call superficial, or cutaneous small-pox. As to remote or derivative diseases, I know of no such thing fairly to be ascribed to the cow-pox; and I have ample means of knowing if such a thing had taken place, as the people of my two parishes, and many in the neighbourhood, are, somehow or other, continually coming under my consideration for medical assistance. My communications of late years have been to Dr. Harvey, according to the directions of the National Establishment; but I have met with no demand for inoculation since February 1820, simply from the absence of any stimulating alarm. I am, dear Sir,

With the highest respect,

Your most obedient humble servant,

T. T. A. REED, Rector of Leckhamstead.

Mr. Reed had, in 1806, printed and distributed a little tract for the encouragement of those who entertained any doubt respecting the efficacy of vaccine inoculation. It contains a brief and conclusive history of the practice of vaccination, and is peculiarly honourable to this benevolent clergyman, who so actively exerted himself.

In connexion with Dr. Jenner's LETTER on the influence of eruptive diseases on the progress of the

vaccine vesicle, I now proceed to mention *another* on
a subject which had occupied his attention for many
years. This, his last published work, came out in
1822. It was entitled, " A Letter to Charles Henry
Parry, M.D. F.R.S. &c. &c. &c. on the Influence of
Artificial Eruptions in certain Diseases incidental to
the Human Body." It is not my intention to enter
much at length into the doctrines contained in this
publication. It was printed in the quarto form, and
extended to about sixty-six pages. It was somewhat remarkable that Jenner's first published observations referred to the preparation of the tartar
emetic, and that his last were directed to the agency
of this medicine in curing disease. This subject
occupied his mind very intensely for some time
before his death, and it affords (I think) a proof that
he had permitted his favourite method of reasoning
by analogy to carry him farther than perhaps was
wise. He threw out his opinions, however, with
great modesty, putting them forward rather as questions or speculations than as doctrines or dogmas.
It contained, nevertheless, many excellent practical
facts and observations. It would open up a large
field of physiological and pathological remark were I
to attempt to give an account of the views which he
entertained. For those who are desirous of information on these points, I must refer to the publication itself. It is interesting, nevertheless, to find
Jenner at the close of a long life busied with topics
calculated to advance our knowledge of diseases,
and add to our means of removing them. He carried on an extensive correspondence with his pro-

fessional friends ; he collected cases illustrative of his opinions ; and altogether pursued his work with as much ardour and earnestness, as if he had been commencing his career, and never had effected any thing for mankind.*

Not many months after the publication of this paper, Dr. Parry lost his father ; Jenner was one of his oldest and most attached friends. He went to Bath to attend his funeral, which took place about the middle of January 1822. " Poor Parry ! " he observes. " I have just returned from Bath, where I went to attend his remains to the silent tomb. The manifestations of regard and affection exhibited by all ranks from Sion Hill to the Abbey, bore unequivocal testimony to his worth and talents."

The Observations on the Migrations of Birds were read before the Royal Society on November 27, 1823. They were presented to Sir Humphrey Davy by the Rev. G. C. Jenner, who, to use his own words, "had the peculiar happiness to accompany his uncle in most of the investigations of the phenomena of migration." " Had it pleased Providence to have spared him a little longer, he might probably have corrected some inaccuracies in the style and order of his paper, that may now, perhaps, appear conspicuous to the reader, but which I did not conceive myself justified in attempting."

It was not the intention of the paper to give a general history of the migration of birds, but rather

* During the progress of these his last labours, he was assisted by Mr. now Dr. John Fosbroke, for whose success and well-being he always expressed an anxious concern.

to communicate some facts, with respect to the cause which impels the bird, at certain seasons of the year, to quit one country for another. He first proves by well-selected incidents, that birds do migrate; and of course he rebuts that doctrine which ascribes their disappearance to a state of hybernation.

I cannot quit this production without alluding to its character as a literary composition. Though it did not receive the last polish of the author's hand, there are several parts of it which have evidently been finished with great care. Some of the descriptions, indeed, are exceedingly beautiful; and could not have been written by any one who did not unite scientific accuracy with a poetical imagination. The following passage, I feel assured, will fully justify these remarks:

"First, the robin, and not the lark, as has been generally imagined, as soon as twilight has drawn the imperceptible line between night and day, begins his lonely song. How sweetly does this harmonize with the soft dawning of day! He goes on till the twinkling sunbeams begin to tell him his notes no longer accord with the rising scene. Up starts the lark; and with him a variety of sprightly songsters, whose lively notes are in perfect correspondence with the gaiety of the morning. The general warbling continues, with now and then an interruption, for reasons before assigned, by the transient croak of the raven, the screaming of the jay and the swift, or the pert chattering of the daw. The nightingale, unwearied by the vocal exertions of the night, withdraws not proudly by day from his inferiors in song, but joins them in the general harmony. The thrush is wisely placed on the summit of some lofty tree, that its loud and piercing notes may be softened by distance before they reach the ear; while the mellow blackbird seeks the inferior branches. Should the sun, having been eclipsed with a cloud, shine forth with fresh effulgence, how frequently we see the goldfinch perch on some blossomed bough, and hear his song poured forth in a strain peculiarly energetic, much more sonorous and lively now than at any other time; while the sun full shining on his beautiful plumes, displays his golden wings and crimson chest to charming advantage. The notes of the cuckoo blend with this cheering concert in a perfectly pleasing manner, and for a short time are highly grateful to the ear; but sweet as this singular song is, it would tire by its uniformity were it not given in so transient a manner. At length, the evening advances, the performers gradually retire, and the concert softly dies away The sun is seen no more. The robin again sets up his twilight song, till the still more serene hour of night sends him to the bower to rest: and now, to close the scene in full and perfect harmony, no sooner is the voice of the robin hushed, and night again spreads a gloom over the horizon, than the owl sends forth his slow and solemn tones. They are more than plaintive and less than melancholy; and tend to inspire the imagination with a train of contemplations well adapted to the serious hour. Thus we see that birds, the subject of my present inquiry, have no inconsiderable share in harmonising some of the most beautiful and interesting scenes in nature."

CHAPTER IX.

DOMESTIC HABITS AND PERSONAL CHARACTER—DEATH AND
FUNERAL.

It is a source of great satisfaction to all who were
acquainted with Jenner, to reflect on the moderation
and wisdom with which he conducted himself during
the whole of the painful controversy that arose out
of his discovery. He had a mind of peculiar delicacy, and he regarded his fair fame with trembling
jealousy ; but he knew that his researches had been
conducted with perfect fairness, that he had no
personal or selfish feelings to gratify, and that a
magnanimous and virtuous desire to render his
knowledge a source of advantage to his fellow-creatures, guided all his actions. Though often stung
by the severe and unmerited reproaches of open
enemies, and wounded still more by the desertion
of his familiar friends ; yet, under these circumstances, of all others most trying to the spirit of man,
and most likely to stir up within him the hot and

vehement passions of our nature, he was enabled to preserve himself calm and unruffled. The attempts to injure his reputation, to impeach his moral character, or to interfere with that distinction and reward which his country conferred upon him, called forth no angry expressions. His composure and his forgiving disposition under trials of this kind cannot be too much admired, or too often held up for example.

He was blest with a helpmate, who not only herself experienced " peace amidst billows," but was permitted to extend the influence of that spirit which sustained and comforted her to all around. When vexed and harassed, he knew where to seek refuge; he knew where dwelt love and truth; he knew where to flee from unjust judgments, and to make his appeal where it never was made in vain. I do not mean to affirm that Dr. Jenner at all times, or during the whole course of his life, participated in the deep and inexhaustible sources of strength and consolation which so manifestly nourished the heart and guided the understanding of his partner; but I should act unjustly by that principle which directed her, were I not to avow it as my firm conviction that it is to her devout and holy life, and her meek and firm and consistent conduct that we are, in some measure, enabled to dwell with so much pleasure upon the memory of her husband. To such an influence it must be ascribed that he was kept so free from the taint of human passions and imperfections when they were most likely to be excited; that he shewed so much genuine modesty, so cordial a desire to do good to his enemies, and all those other qualities that grow not from an earthly root.

I remember, when discussing with him certain questions touching the condition of man in this life, and dwelling upon his hopes, his fears, his pains, and his joys, and coming to the conclusions which merely human reason discloses to us; and when dwelling on the deformity of the heart, our blindness, our ignorance, the evils connected with our physical structure, our crimes, our calamities, and our unfathomable capacity both for suffering and for enjoyment; he observed, Mrs. Jenner can explain all these things : they cause no difficulties to her.

The observations which this remark suggested were not pursued at the time, but were often recurred to on subsequent occasions. As he approached nearer to his own end, his conversations with myself were generally more or less tinged with such views as occur to a serious mind when contemplating the handiwork of the Creator. In all the confusion and disorder which appears in the physical world, and in all the anomalies and errors which deface the moral, he saw convincing demonstration that He who formed all things out of nothing still wields and guides the machinery of his mighty creation.

In his early days he certainly, I fear, had fallen into that error too common among men who have been much occupied in the pursuit of mere human knowledge—he did not clearly discern the difference between the things which are made known to us through the medium of our senses and our reasoning faculties, and those which come to us with higher claims, both upon our affections and our understanding. Manifold evils have arisen from this cause. Great and rapid as has been the advancement of every science, it is to be feared that the state of mind of the enquirers has not always accorded with true wisdom; that they have sometimes mistaken the farthest end and aim of all knowledge, as well as the best means of attaining it.

One of the most remarkable features in Jenner's character, when treating of questions of a moral or scientific nature, was a devout expression of his consciousness of the omnipresence of the Deity. He believed that this great truth was too much overlooked in our systems of education ; that it ought to be constantly impressed upon the youthful heart, and that the obligations which it implied, as well as the inward truth and purity which it required, should be rendered more familiar to all. Mrs. Jenner was constantly occupied in teaching these lessons to the poor around her in schools, which she established for the purpose of affording a scriptural education. He, building upon this foundation, wished to add instruction of a more practical description, deduced from their daily experience, and illustrated by a reference to those works of wisdom and beauty which the universe supplies. He always contended that some aid of this kind was necessary to impress completely upon the character of the lower ranks those maxims which they derived from their teachers. He had other views, too, in recommending such a plan ; he thought that the lot of the poor might be ameliorated, and many sources of amusement and information laid open to them which they are at present deprived of; that the flowers of the field and the wonders of the animal creation might supply them with subjects of useful knowledge and pious meditation.

The state of Mrs. Jenner's health often required the unremitting and anxious care of her husband. She had been long threatened with a pulmonary complaint of a serious character, and was besides affected with another disorder of a painful and distressing nature. For years before her death she was chiefly confined to her own apartments. The tenderness and delicacy with which Jenner superintended the arrangement of every thing that could be thought of for her comfort, the administration of her medicine and the preparation of her food, (which a difficulty of deglutition rendered necessary,) all indicated the warmest attachment and the kindest feelings. The unaffected cheerfulness and thankfulness with which, amid her pains, she received such offices, was truly instructive. This temper was conspicuous at all times—in the days of comparative health as well as at the hour of death.

Dr. Jenner's personal appearance to a stranger at first sight was not very striking; but it was impossible to observe him, even for a few moments, without discovering those peculiarities which distinguished him from all others. This individuality became more remarkable the more he was known; and all the friends who watched him longest, and have seen most of his mind and of his conduct, with one voice declare, that there was a something about him which they never witnessed in any other man. The first things that a stranger would remark were the gentleness, the simplicity, the artlessness of his manner. There was a total absence of all ostentation or display ; so much so, that in the ordinary intercourse of society he appeared as a person who had no claims to notice. He was perfectly unreserved, and free from all guile. He carried his heart and his mind so openly, so undisguisedly, that all might read them. You could not converse with him, you could not enter his house nor his study, without seeing what sort of man dwelt there.

His professional avocations and the nature of his pursuits obliged him to conduct his inquiries in a desultory way. At no period of his life could he give himself up to continued or protracted attention to one object : there was, nevertheless, a steadiness in working out his researches, amid all the breaks and interruptions which he met with, that can only belong to minds constituted as his was.

The objects of his studies generally lay scattered around him ; and, as he used often to say himself, seemingly in chaotic confusion. Fossils, and other specimens of natural history, anatomical preparations, books, papers, letters—all presented themselves in strange disorder ; but every article bore the impress of the genius that presided there. The fossils were marked by small pieces of paper pasted on them, having their names and the places where they were found inscribed in his own plain and distinct handwriting. His materials for thought and conversation were thus constantly before him; and a visitor, on entering his apartment, would find in abundance traces of all his private occupations. He seemed to have no secrets of any kind: and, notwithstanding a long experience with the world, he acted to the last as if all mankind were trustworthy, and free from selfishness as himself. He had a working head, being never idle, and accumulated a great store of original observations. These treasures he imparted most generously and liberally. Indeed his chief pleasure seemed to be in pouring out the ample riches of his mind to every one who enjoyed his acquaintance. He had often reason to lament this unbounded confidence; but such ungrateful returns neither chilled his ardour nor ruffled his temper.

In the success of his researches he was a proof of the felicity with which an humble spirit is rewarded who proceeds to investigate that great field of knowledge, "which has been passed to man by so large a charter from God." He was not arrogant nor confident, but was contented to learn with all the docility of a child.

In prosecuting his investigations into unexplored regions, analogy was his favourite guide. This method is characteristic of all original minds; and although it is often carried too far, it has been, when duly and cautiously followed, the parent of some of the greatest inventions. To it we are, in great degree, indebted for the discovery of the properties of the Variolæ Vaccinæ. In comparing that affection with the disease it was intended to counteract, he gained his chief knowledge, and was ultimately enabled so to establish their points of difference, as to render the information he had acquired of the highest practical utility. For this sound and sagacious mode of reasoning he was principally indebted to his own vigorous understanding. He had not derived much aid from those helps which wise men have devised for keeping the intellect in its proper course while searching for truth ; and to this cause we may probably ascribe some of the errors into which, it must be confessed, he occasionally fell. His analogies were sometimes hurried forward on the wings of imagination ; and, of course, were not always accurate or conclusive. His language, too, on scientific subjects, though for the most part remarkably simple and precise, was, on some occasions, of too figurative a cast. This rich and flowery garb often seemed to overlay sterling treasure, and by those who could not penetrate below the surface he has been deemed rather visionary.*   But, truly, it

* I am glad to have a confirmation of the above remarks from the pen of his illustrious friend the late Sir Humphrey Davy.

" I remember," says Sir Humphrey, " in 1809, having had a long conversation with the late Dr. Jenner on the habits of animals. He was original and ingenious, but I think was sometimes carried too far by the remoteness of his analogies. We were discussing the possibility of the uses of earthworms to man.   I was more disposed to consider the dunghill and putrefaction as useful to the worm, rather than the worm as an agent important to man in the economy of nature ; but Dr. Jenner would not allow my reason. He said the earthworms, particularly about the time of the vernal equinox, were much under and along the was a misapprehension.   His comparisons were often most happy and appropriate; and of this I can scarcely refer to a better example than a reply that he made to Charles Fox, which will be found in a subsequent page.

As faithfulness is one of the first qualities of a biographer, I may here notice what I take to be a fair instance of his analogical reasoning pushed to an undue extent.   In his work on artificial eruptions, most of his anticipations of the benefits to be obtained from them in different diseases were evidently deduced from a supposed affinity between those that are artificial and those that are natural.   Though the analogy is correct in some respects, it certainly does not hold true to the extent which he imagined, nor are the benefits, great as they have been proved to be, altogether of that nature which he had conjectured.

In witnessing the variety of external things, and in marking their properties, he seems to have possessed a mind much allied to the pure and unsophisticated character of some of our old English worthies; and were I to attempt to find associates with whom

surface of our moist meadow lands; and wherever they move, they leave a train of mucus behind them, which becomes manure to the plant.   In this respect they act, as the slug does, in furnishing materials for food to the vegetable kingdom; and under the surface, they break the stiff clods in pieces, and finally divide the soil.

"They feed likewise entirely on inorganic matter, and are rather the scavengers than the tyrants of the vegetable system." (See Davy's Life of Davy, vol. ii. p. 389).

he would in an especial manner have assimilated, I think I should seek to link him in triple union with honest Isaak Walton and the pious and engaging Evelyn. When I read some of the descriptions of the former, but especially that lovely and heart-stirring passage where the mention of the doubling and redoubling notes of the nightingale is so finely employed to arouse men to the beauties of this crea-tion, and to point their hopes ɔ another, I have often been reminded of delineations of a similar kind in the writings of Jenner. That passage towards the conclusion of his paper on the migration of birds, which has been given at page 279, will serve as an illustration.

His latter days were occasionally gladdened by the studies and pursuits of his youthful years. Geo-logy, which had made such rapid strides since he commenced his enquiries, continued to interest him to the last. He had several visits from the learned and distinguished Professor Buckland, with whom, and the Rev. R Halifax of Standish, he examined the trap rocks at Micklewood, and the corals and agates at Woodford. He also, with the assistance of Mr Henry Shrapnell, arranged his own specimens of natural history.

His domestic happiness after the death of Mrs. Jenner was necessarily much impaired. Another bereavement, though of a different kind, which oc-curred not long before his death, rendered his state still more desolate. His only daughter Catherine was on the 7th of August 1822, married to John

Yeend Bedford, Esq. of Southbank, Edgbaston, near
Birmingham. The day that this event took place,
he sent me the following note. " Pray don't desert
this forlorn cottage, but come sometimes, and chase
away my melancholy hours.

" With best affections,
" EDW. JENNER.

" *Chantry Cottage, 7th August,* 1822."

The union had taken place with his entire approbation. This amiable lady was delivered of a daughter on the 1st of August 1833, and expired on the 5th
of the same month. This only offspring of the mar-
riage is named Catherine Sarah Jenner.

His habits were in perfect accordance with the unaffected simplicity of his mind; and never, probably,
did there exist an individual to whom the pomp and
ceremony, which are so pleasing to many, would have
been more burdensome. Unrestrained by the formality and reserve of artificial society, he loved to
enjoy that freedom, in his intercourse with his friends,
which was always gratifying to them, and congenial
to his own taste.

In his latter years he was not a very early riser;
but he always spent some part of his time in his
study before he appeared at the breakfast table.
When in London and at Cheltenham, he generally
assembled his scientific and literary friends around
him at this hour. Some came for the pleasure of
his conversation; some to receive instruction in the
history and practice of vaccination. In the country,
where his guests were generally his own immediate

connexions or his intimate friends, the originality of his character came out in the most engaging manner. He almost always brought some intellectual offering to the morning repast. A new fact in natural history, a fossil, or some of the results of his meditations, supplied materials for conversation ; but, in default of these, he would produce an epigram, or a fugitive jeu d'esprit ; and did not disdain even a pun when it came in his way. His mirth and gaiety, except when under the pressure of domestic calamity or bodily illness, never long forsook him ; and even in his old age, the facility with which he adapted his conversation and his manners to the most juvenile of his associates was truly interesting. To have seen and heard him at such times, one could hardly believe that he was advanced in years, or that these years had been crowded with events so important.

Though thus kind, and free, and familiar, there was nothing of levity in his deportment ; and, when occasion required, he could well sustain the dignity of his name and station. In the drawing-room at St. James's he chanced to overhear a noble lord, who was high in office, mentioning his name, and repeating the idle calumny which had been propagated concerning his own want of confidence in vaccination, in consequence of his acting as has been already stated in the case of his son Robert. He, with the greatest promptitude and decision, refuted the charge and abashed the reporter. His person was not known to the noble lord, but with entire composure he advanced to his lordship, and looking fully in his face, calmly observed, " I am Dr. Jenner." The effect of this well-timed rebuke was instantaneous. The noble lord, though " made of .sterner stuff " than most men, immediately retreated, and left Jenner in possession of the field.

As he knew how to comport himself with men of elevated rank, he could condescend to his inferiors in the most benevolent and gracious manner. He loved to visit and to converse with them ; to observe their domestic habits, and the little peculiarities in language or demeanour which different districts exhibit; but he especially delighted in discovering any traces of originality, any indi_ cations of that *vivida vis animi* which might with a little help enable the possessor to emerge from his humble station. Young Worgan, who became tutor to his eldest son, was fostered by him in this way, and there are many others still living who have equally partaken of his encouragement and of his bounty.

He was particularly fond of conversing with people in the lower walks of life who were of a religious character. I know one venerable individual of this kind, who has likewise given proofs of very considerable musical genius. Though compelled to labour at an humble trade, and little indebted to education, poor Thomas Cam contrived, while living in a secluded hamlet, to acquire such a knowledge of the theory of music as to be able to compose pieces of considerable length, and adapted to a great variety of instruments, some of which he had never seen or heard. Jenner on one occasion brought him to the

Music-meeting at Gloucester. There he witnessed
an orchestra more varied and complete than any
he had ever before contemplated. He listened
with extraordinary satisfaction; and when Jenner
asked him if he was not astonished at the strange
concord of sweet sounds issuing from a number of
instruments new to him, " Oh!" said he, " I *knowed*
how it would all be." Several of this poor man's
compositions have been printed; and I am told by
good judges that, considering his opportunities, they
are very astonishing.

Every indication of talent or genius, in whatever
situation found, was sure to gain his notice and consideration. I remember to have seen him, a short
time before his death, listening with great attention
to the demonstrations of a very humble lecturer on
astronomy. The Doctor had collected all his young
friends in Berkeley about him in his own house, and
the lecturer, though very insufficiently provided with
instruments, and little beholden to any thing but his
own exertions for his knowledge, was, nevertheless,
animated and ardent. His apparatus and his draw-
ings were all constructed by himself; and, rude
though they were, they fixed the attention of the
younger part of the audience, and in so doing amply
gratified Jenner. It ought at the same time to be
mentioned, that neither Dr. Jenner's previous education nor his habits gave him a relish for any of the
branches of pure science. He seemed to have a peculiar horror of arithmetical questions. He was often jocular on this defect in his nature; and I believe he frequently paid severely for it; as he would rather attend

to any thing than pounds, shillings, and pence. A neighbour was once expending a great many words to draw his attention to some affairs of this kind. He expressed himself perfectly satisfied; but not so his neighbour. He continued to dwell upon the different items till Jenner's patience became exhausted; and he exclaimed that he would rather look for an hour at a mite through a microscope than have his time taken up with such things.

Whether in the country or in town, his eye was constantly in search of subjects for observation. He seldom or never passed a butcher's shop without a peep at its contents; because he often found something to illustrate his views of comparative anatomy and pathology. He generally carried a large pocket-book with him; and recorded his thoughts as they occurred. He very often also adopted another practice, namely, that of writing his reflections on detached scraps and fragments of paper; and many, consequently, have either perished or been rendered useless for want of connexion: these " disjecta membra " being not very susceptible of arrangement or combination by any other than the mind which produced them.

Though the general cast of his character exhibited a happy union of great solemnity and seriousness with extraordinary playfulness, amounting at times even to the height of mirth and jocularity; yet no one ever found these latter qualities misplaced, or obtruding themselves unseasonably. Almost all the great incidents of his life tended rather to suppress them, and to keep them in the shade. In the early part of this work it has been shewn, when meditating on the grand results of his vaccine experiments, how devout were his feelings. Towards the close of his life many incidents argue the increasing power of that principle. He frequently expressed his regret that mankind were so little alive to the value of vaccination. Among the last words that he addressed to me, not many days before his fatal seizure, he used this remarkable expression: " I am not surprised that men are not thankful to me; but I wonder that they are not grateful to God for the good which he has made me the instrument of conveying to my fellow creatures." He had a great reverence for the Scriptures ; and when he presented copies of them to his god-children or others, they never went from his hands without some inscription, declaratory of his veneration : one such, I subjoin.

" To Augusta Bertie Parry, with the best wishes and affections of her god-father, Edward Jenner ; who most devoutly hopes, as this is the best book that ever was written, she will give it not only the first place in her library, but convince those who love her dearly, that it occupies the first place in her heart."

I find some fragments of prayers strongly expressive of deep and humble submission to the divine will. One of them, apparently written under affliction, concludes in this strain :—" And may those sacred truths, revealed by him who did condescend to assume a human form, and appear among men upon the earth, be so engrafted in my mind, that I may never lose sight of these thy divine mercies ; and thus, by my faith and practice, when it may please thee to send my body to the grave, may my imperishable soul be received into thy habitations of eternal glory."

Dr. Jenner purchased the house which he inhabited at Berkeley from a family of the name of Weston. It was called "the Chantry," from having, in former times, been in the possession of certain monks. It is contiguous to the churchyard of Berkeley; and the tower of the church, which, as is sometimes the case, is disjoined from the rest of the building, overhangs the southern boundary of the shrubbery. This tower is now nearly covered with a vigorous ivy plant, which on two sides has mantled to its summit. Jenner plucked the root from which it sprang from the tomb of Strongbow * at Tintern Abbey. He carried it with him to Berkeley, and planted it, observing, "who knows but this little scion will one day encircle our goodly tower?"

The tower, in its green and rich livery, is now a beautiful object, and harmonizes with the shrubs and trees which tastefully adorn his little domain. One tree (a willow) is conspicuous, as well for its light and beautiful foliage, as from its having been a great favourite with him. He particularly loved to endear to himself all objects with which he was familiar, by associating them with some incident calculated to mark past events in his personal history. This very tree was so distinguished. His eldest son, to whom he was devotedly attached, was, in consequence of severe illness, obliged to lose blood,

---

* Richard Earl of Clare, who died in 1176.

which Jenner deposited with his own hands at the root of this tree.

Towards the southern extremity of the lawn, and shaded by the thick screen of evergreens, is a small rustic apartment. Here in the summer mornings Jenner used to receive his poor neighbours who came for the purpose of vaccination. In this humble fane more wonders were wrought than in all the splendid temples of Æsculapius. It was constructed by the Rev. Mr. Ferryman. The knotted and gnarled oak, with huge fragments of the roots or branches of other forest trees, arranged with much taste, enabled him to give to an extremely artificial structure the style and character of a natural production. The monarch of the woods, shorn of his glory, and dying internally, but still holding his attachments to his parent earth, sometimes exhibits an arched cavity, furnishing hints for such structures. If the reader has ever examined the Greendale oak, as described by the excellent Evelyn, he will easily comprehend the idea I wish to convey. It was not merely in copying the vegetable world, either in its soft and lovely character, or in its bold and picturesque effects, that Mr. Ferryman shewed the accuracy of his eye and the correctness of his taste. He could imitate with equal fidelity the abrupt and varied form of a rocky surface, and could so dispose of the massy fragments riven from an adjacent quarry, as if they had been fixed by some great convulsion on the surface of the earth, not by the puny efforts of human hands; but Mr. Ferryman had a frame that well seconded the conceptions of his mind. There was nothing little in any of his conceptions. They all resembled the operations of nature in her firmest and most decided displays; and with his own hands he would labour in executing his designs with irresistible energy.

In the style of ornamental improvement, which within the last half century has done so much to augment the natural beauties of England, Mr. Ferryman was quite unrivalled. He followed one guide; and so admirably did he adapt his alterations to the situation and character of the surrounding objects, that they seemed rather like parts of one original design than artificial adjuncts. I do not know that he ever read the elegant work of the late accomplished Sir Uvedale Price, Bart. but there is a relationship between their conceptions, and a truth in their practical elucidations, which stamps them as brothers in the same family of genius.

I would hope that the great design of writings of this description has in some degree been attained; that the form and likeness of the mind has, to a certain extent, been preserved. The incidents which have been recorded, will afford to every one the means of tracing the features of the character. I am fully conscious, nevertheless, that the portrait but feebly delineates the merits of the original. It is not my intention to attempt to increase the effect by artificial colouring, or exaggerated representations; but there are some characteristic traits that I have yet to mention.

The discovery of vaccination, though pregnant with consequences, calculated from their magnitude to dazzle and bewilder the strongest intellect, was

ushered into the world with singular modesty and
humility. It soon, however, began to expand; and
when the opposition arose, its value became more
apparent, and its power and its virtues were demonstrated by the very objections that were brought
against it. In these respects it resembled truth of
a different description, which becomes more resplend-
ent and glorious, the more it is tried by controversy
or persecution.

But Dr. Jenner was not only humble in all that
concerned this, the greatest incident of his life; he
continued so after success had crowned his labours,
and after applause greater than most men can bear
had been bestowed upon him. This most estimable
quality was visible at all times; but it was particularly conspicuous when he was living in familiar intercourse with the inhabitants of his native village.
If the reader could in imagination accompany me
with him to the dwellings of the poor, and see him
kindly and heartily inquiring into their wants, and
entering into all the little details of their domestic
economy; or if he could have witnessed him listening with perfect patience and good humour to the
history of their maladies, he would have seen an
engaging instance of untiring benevolence. He
never was unwilling to receive any one, however
unseasonable the time may have been. Such were
his habits, even to the latest period of his life. I
scarcely know any part of his character that was more
worthy of imitation and unqualified respect than that
to which I have alluded. I have never seen any
person in any station of life in whom it was equally

manifest; and when it is remembered that he was well "stricken in years;" that he had been a most indefatigable and successful labourer in the cause of humanity; and that he might have sought for a season of repose, and the uncontrolled disposal of his own time, the sacrifices which he made are the more to be valued. In the active and unostentatious exercise of kindness and charity he spent his days; and he seemed ever to feel that he was one of those " qui se natos ad homines juvandos, tutandos, conservandos arbitrantur."

His kindness and condescension to the poor was equalled by his most considerate respect and regard to the feelings and character of the humblest of his professional brethren. I have often been struck with the total absence of every thing that could bear the semblance of loftiness of demeanour. Few men were more entitled to deliver their sentiments in a confident or authoritative tone; but his whole deportment was opposed to every thing of that description, and he did not hesitate to seek knowledge from persons in all respects his inferiors. All his younger brethren who have ever had the happiness to meet him in practice, must have been deeply impressed with this part of his character.

He had both an inquisitive and an original mind; and it was always open to instruction, from whatever quarter it came. He seldom failed, either when writing to his professional brethren, or when conversing with them, to start some subject for their consideration. I have known him often dictate to his young friends problems in physiology, pathology, or natural history, for their investigation; at the same time giving them some important information which he had previously ascertained by his own inquiries. Some of the pathological questions, which it has been my lot to discuss, originated in this way, and were prosecuted with his fostering help.

I am satisfied, as I have already observed *, that the overwhelming duties connected with vaccination have in some measure obstructed and obscured the reputation which is his due as a scientific physician. Very many of the subjects which are now occupying the attention of the profession, and which have led to valuable practical results, were fully developed in his mind; and had he been permitted to have brought them before the public, he would have earned a well-deserved accession to his fame.

He had a strong and just feeling of consideration for the many hardships endured by medical men, particularly in country districts. A dreary ride over a bleak and wintry road in the middle of the night, and a cold and comfortless reception in the abode of sickness and poverty, with nothing for the rider or his horse; or, what is worse, an urgent and impatient summons to a more wealthy abode, where all consideration is centred in one point, and where no provision is made for the unhappy son of Æsculapius; these things he felt so strongly, that he used to illustrate his sentiments in a jocular manner, by saying, that medical gentlemen should follow the example of tradesmen, and endeavour to bring their employers to a sense of justice by " *a general strike.*"

* Vol. i. p. 120.

The infirmity of our nature leads us too often to draw inferences from circumstances that are very fallacious. We are too apt to attach ideas to persons and things that do not necessarily belong to them, and to imagine that whatever does not correspond with our preconceived opinions must be erroneous. Science is supposed to flourish only in certain regions; and new or unexpected information is sure to meet with the reception which is due only to unfounded pretension. Perhaps no man of his day had more cause to lament this bias than Jenner. He used often to say, " I believe there are many individuals in our profession who estimate a man's intellect by the size of the place in which he lives. Of course, I must be a very small person, seeing that our good town of Berkeley cuts such a sorry figure. I, to be sure, have been in authority; and my office of mayor may have given me some consequence among the townsfolk; but I have often found my opinions resisted by my professional brethren, when my influence, perhaps, ought to have been greater than in my civic capacity." In this manner he used often to laugh when he alluded to the reception that many of his opinions encountered. At other times his feelings were of a more serious character, especially when he lamented the great injury that was done to the cause of vaccination by an unwillingness on the part of many influential persons to examine what he had said, or to give credit to his statements. This reluctance may, and often does, impede the cause of truth; and deprive deserving men of their just meed of credit and approbation.

During his residence at Berkeley he acted frequently as a magistrate. I found him one day sitting with a brother justice in a narrow, dark, tobacco-flavoured room, listening to parish business of various sorts. The door was surrounded by a scolding, brawling mob. A fat overseer of the poor was endeavouring to moderate their noise ; but they neither heeded his authority nor that of their worships. There were women swearing illegitimate children, others swearing the peace against drunken husbands, and able-bodied men demanding parish relief to make up the deficiency in their wages. The scene altogether was really curious ; and when I considered who was one of the chief actors, and saw the effect which the mal-administration of a well-intended statute produced, I experienced sensations which would have been altogether sorrowful had there not been something irresistibly ludicrous in many of the minor details of the picture. He said to me, " is not this too bad ? I am the only acting magistrate in this place, and I am really harassed to death. I want the Lord Lieutenant to give me an assistant ; and I have applied for my nephew, but without success."

On this visit he shewed me the hide of the cow that afforded the matter which infected Sarah Nelmes : and from which source he derived the virus that produced the disease in his first patient Phipps. The hide hung in the coach-house : he said, " What shall I do with it ?" I replied, " send it to the British Museum." The cow had been turned out to end her days peaceably at Bradstone, a farm near Berkeley.

He talked of the first effects of his discovery on some of his sapient townsfolk. One lady, of no mean influence among them, met him soon after the publication of his Inquiry. She accosted him in this form, and in the true Gloucestershire dialect. " So, your book is out at last. Well! I can tell you that there be'ant a copy sold in our town; nor sha'n't neither, if I can help it." On another occasion, the same notable dame having heard some rumours of failures in vaccination, came up to the doctor with great eagerness, and said, " Shan't us have a general inoculation now?"*

Both these anecdotes he used to relate in perfect good humour.

On another occasion, when travelling with him towards Rockhampton, the residence of his nephew Dr. Davies, he observed, " it was among these shady and tangled lanes that I first got my taste for natural history." A short time afterwards we passed Phipps, his first vaccinated patient. " Oh! there is poor Phipps," he exclaimed, " I wish you could see him; he has been very unwell lately, and I am afraid he has got tubercles in the lungs. He was recently inoculated for small-pox, I believe for the twentieth time, and all without effect."

At a subsequent visit, (Oct. 1818,) I found lying on his table a plan of a cottage. " Oh," said he, " that is for poor Phipps; you remember him: he

---

* *i. e.* small-pox inoculation.

has a miserable place to live in ; I am about to give him another. He has been very ill, but is now ma terially better." This cottage was built, and its little garden laid out and stocked with roses from his own shrubbery, under his personal superintendence.

I may now mention some incidents of a different character. The celebrated Charles James Fox, during a residence at Cheltenham, had frequent intercourse with Jenner. His mind had been a good deal poisoned as to the character of cow-pox by his family physician, Moseley. In his usual playful and engaging manner, he said one day to Jenner, " Pray, Dr. Jenner, tell me of this cow-pox that we have heard so much about :—What is it like ?" " Why, it is exactly like the section of a pearl on a rose-leaf." This comparison, which is not less remark-able for its accuracy than for its poetic beauty, struck Mr. Fox very forcibly. He laughed heartily, and praised the simile.

It has been seen, that notwithstanding the per-sonal influence that Dr. Jenner had with foreign states, he had next to none at home. He never succeeded in procuring an appointment for any of his relatives or friends. He mentioned that all his attempts to get a living for his nephew George had failed, though addressed to quarters where they might, without presumption, have been expected to have met with attention and success. This neglect hurt him deeply. He once said to me, " This ought to be known. You must give them a hard one ; and I will find an eagle's quill and whet the nib for you." *

* His favourite eagle had just died.

I never saw him more happy than in spending some days with Dr. Baillie at Duntisbourne, near Cirencester, in the summer of 1820. He had much recovered from the impression left by the death of Mrs. Jenner; and all the recollections of his youth, his intercourse with Mr. Hunter, together with many of the remarkable incidents which were connected with his own life, formed animating themes for conversation. The scenes around them, also, in the vicinity of the place (Cirencester) where he had first gone to school, and where he used to grope for fossils in the oolitic formation, supplied him with many associations of long-past years. I spent one of the days with them on this occasion. They passed their time in the free and unreserved interchange of their thoughts and their experience.

It was cheering to see the great London physician mounted on his little white horse, riding up and down the precipitous banks in the vicinity of his house, or trotting through the green lanes, and opening the gates, just after the manner of any Cotswold squire. Nothing could exceed the relish of Baillie for the ease and liberty and leisure of a country life, when he first escaped from the toil and effort and excitement of his professional duties in London. Duntisbourne stands in rather a picturesque situation; the house overhangs a deep wooded dell, and is fronted on the opposite bank by the church and hamlet of Edgworth. The ramifications of this dell are intricate and beautiful; but there was little else in the doctor's vicinity to gratify the eye. Every thing wore an aspect of cheerfulness to him; and whether he was traversing the bleak summit of the Cotswolds or taking his pastime in the more cultivated domains of Pimbery or Oakeley,* he was equally happy and equally buoyant.

Jenner's intimacy with his uncle, John Hunter, had established a sort of family connexion, which the subsequent events in the life of each ripened into the most cordial regard and attachment. Jenner, in the meadows of Gloucestershire, had achieved a discovery which at once raised him to the highest point of professional reputation. Baillie, in the metropolis, was running a career of honour and usefulness derived from the soundest knowledge, and adorned with all the virtues that can render such knowledge most estimable. He was among the first to appreciate correctly the value of vaccination, and to stand forward to vindicate its character when it was traduced and vilified; and as long as he lived, he lent all the influence of his name and authority to the practice. His own plain, direct, and honest heart, taught him promptly to discover and appreciate kindred qualities in others.

I had the happiness of seeing them both together again in my own house on the 30th of August, 1821, where they spent the night.† Jenner in the interval

---

* Belonging to Earl Bathurst.

† During the same autumn Jenner also visited his friend John Phillimore Hicks, at Eastington, and his nephew, Edward Davies, at Ebley. While at the latter place he sat for his picture to Mr. Hobday. An engraving has been made from it, of the same size as that of the celebrated print of John

had sustained a serious illness. He was in pretty good spirits; and his ardour for knowledge was unabated. I remember he brought in his pocket some fossils, and one of the vertebræ of the back of a horse, to show the nature of the change which takes place in that disease called string halt.

I fondly hoped from the vigour which they both then exhibited, that their lives might have been spared for many years; but before two were over, it was my misfortune to see them both laid in their graves.

I have on former occasions mentioned slight illnesses with which Jenner had been affected at different periods. They all more or less pointed to that sudden and fatal seizure which ultimately extinguished life.

One attack of a very alarming nature occurred on the 6th of August, 1820. He was walking in the garden, and became suddenly faint and giddy; he sunk to the ground, and his hat dropped off. How long he remained in this state could not be ascertained, as he contrived ultimately to get into the house, where he was found in a somewhat confused state; and his clothes covered with earth. The hat was picked up afterwards near the place where he had fallen. He was put to bed, and immediately visited by his nephew and Mr. Henry Shrapnell. They promptly administered suit-

Hunter. Sharpe commenced the work, and was anxious to make it a worthy companion to his master-piece, which I have just named; but he died before it was finished, and it was completed by Skelton.

able remedies, and the alarming symptoms were removed. An express had been sent to me, but I could not reach Berkeley till two A. M. He was then asleep, and I did not, of course, disturb him. Next morning I had the satisfaction of finding, that, though the attack had been threatening, it had not left any permanent traces of its nature. There was no paralysis, no confusion, no indication of serious mischief having been done to the brain. He was, however, depressed and thoughtful, as became one who had been saved from great peril. Death and its consequences formed an interesting part of our conversation, and his mind on that subject was tranquil and firm. He recurred to the loss of his dear wife; remembered her patience and resignation; and though disquieted a little about some matters of a temporal nature, I could not help rejoicing to find him in a frame of mind so placid and satisfactory.

I saw him again on the 8th, when I found with him his old and valued friend Dr. Worthington. He was still in bed. He received me with great emotion; and shewed that he felt deeply the effects of his illness. I have seldom seen him more moved than he was on this occasion; and I observed plainly, that though the golden bowl was not broken, there was a slight loosening of the silver chord.

He gradually recovered; but the disorder left some distressing results. He became remarkably sensitive to external impressions; but most of all to sounds of a certain description. Those that were

dull and obtuse he little regarded; but the sharp,
harsh click, for instance, of a knife upon a plate, produced an effect as if he had had an electric shock
sent through his frame.

Of course, this painful state of being much marred
his happiness. He could scarcely encounter any of
the most common occurrences of life without being
exposed to great suffering; and the consequence
was, that he could not at times refrain from expressing a slight degree of impatience or irritability under
such inflictions.

His mental efforts were necessarily checked for a
season by the illness above noticed. His brain had
certainly been over-worked; and repose and absence
from exciting objects was indispensably necessary
for him. This state never suited him well, and he
could not long be made to submit to it.

In a few months he was deeply engaged in all his
former occupations; and by the publications mentioned in the preceding chapter, proved that the
energy and activity of his intellect had suffered
no diminution. His attention to the great cause
of vaccination was unremitting; and the letters
and documents which are elsewhere printed will
shew that his sentiments were judiciously and consistently maintained.

His parting statement on this subject must have been
written a very few days before he expired. I found it
on the back of a letter, the post mark on which gives the
date January 14th, 1823. He was taken ill on the 25th,
and died on the 26th of that month. It is not known
that he wrote again on this subject; but be this as

it may, nothing could be more solemn, whether we consider the time or the expression of this his final judgment. " MY OPINION OF VACCINATION IS PRECISELY AS IT WAS WHEN I FIRST PROMULGATED THE DISCOVERY. IT IS NOT IN THE LEAST STRENGTHENED BY ANY EVENT THAT HAS HAPPENED, FOR IT COULD GAIN NO STRENGTH; IT IS NOT IN THE LEAST WEAKENED, FOR IF THE FAILURES YOU SPEAK OF HAD NOT HAPPENED, THE TRUTH OF MY ASSERTIONS RESPECTING THOSE COINCIDENCES WHICH OCCASIONED THEM WOULD NOT HAVE BEEN MADE OUT."

It will afford an instructive lesson to the younger members of his own profession, to witness the undiminished energy with which this venerable man cultivated scientific and professional studies almost to the last hour of his existence. In his comparative retirement at Berkeley, his engagements were of a different nature from what they were at Cheltenham. On looking over his note books, which he kept with considerable regularity, I am astonished that at his advanced age, and with so many momentous affairs pressing upon his mind, he should have been able to chronicle with such perseverance so many observations. One of them carries with it a peculiarly impressive character. It must have been written, at the farthest, on the night immediately preceding his own fatal seizure; but I am inclined to think, from various circumstances, that it was the last conscious effort of his mind, and that the entry was made, perhaps, not many minutes before he himself was afflicted nearly by the same malady that

destroyed his patient.   I give it exactly as it appears
in his diary.

"Mr. Joyner Ellis," (he was a schoolfellow and an
old friend), "from long exposure to severe cold, the
thermometer being many degrees below the freezing
point, was so benumbed, that he was brought home,
after a long journey chiefly in an open carriage, in
a state of paralytic debility : the harmony of all the
vital functions seemed disturbed; and of some he
seemed to be quite deprived.   Being moved, he appeared to feel pain about the chest; and as his
breathing was short and laborious, Mr. H. abstracted
about sixteen ounces of blood from the arm, but without relief.   There was that peculiar effort in breathing
that is not really stertorous, but approaching to it; so
that my prognostic was as unfavourable as it could be."

The events above alluded to, occurred in the evening of Friday the 24th of January 1823.   The gentleman to whom they refer died early on the following morning; and Dr. Jenner was made acquainted
with that fact when he left his bed-room.   From the
tenor of the concluding sentence of the preceding
extract, I am inclined to believe that it was written
after he came to the knowledge of the fatal event.
If so, a very short time must have elapsed before
he himself was nearly in the same condition in which
he had described his patient.

The day preceding this attack, he had not
only, as I have just mentioned, been engaged with
his usual ardour in professional avocations, but had
likewise attended to those acts of mercy and benevolence that the wants of his neighbours and the seve-

rity of the season demanded. He had walked to
the village of Ham, and ordered supplies of fuel to
some of the poor people; and in all respects seemed
as well as he had been for a long time. His previous illnesses certainly indicated a proneness to
that disease which cut him off; but his great tem-
perance and regularity, and his almost complete recovery from their effects, gave reason to hope, that
the disastrous event might have been averted.
He had regained also his cheerfulness and his animation to a very considerable extent; and scarcely
at any period of his existence did he appear to be
more alive to every moral and intellectual enjoyment.

After his walk to Ham, he visited his nephew Stephen in his painting-room. He had for some time
taken great pleasure in attempting to cultivate the
talent which this young man displayed as a drafts-
man. Jenner, in order to rouse him to exertion and
study, was in the habit of employing every incite-
ment that could be devised. Among others, he used
to place short apophthegms and extracts from the
writings of Sir Joshua Reynolds, and other eminent
artists, on the easel or on the walls of the apartment,
so that they might meet the eye of his nephew whenever he began to study.

On the morning of his last visit to him, Stephen
was amusing himself while at his work, by singing a
popular Scotch air. Jenner heard the notes as he
entered the room, and detected an inaccuracy in the
tune: "Oh!" said he, "you are singing; but not in a
right way, let me tell you: this is the manner in
which you ought to do it;" and he then sang a

314 LIFE OF DR. JENNER.

stanza or two. The weather was remarkably cold at
the time, and he went down stairs and brought up
some coals himself, and likewise a seat. While he
was thus occupied, a gentleman came in ; and he
observed, in his jocular and kind way, " You see
Stephen has got a servant ; " and then carried on
the conversation on ordinary topics.

In the evening of this day he visited the gentleman whose case I have recently extracted from his
note-book. Next morning he arose as usual, and
came down stairs to his library. As he did not appear at breakfast, the servant was sent to ascertain
the cause. On entering the room, he found that he
had sunk from the couch on the floor ; and that he
was lying in a state of insensibility. His nephew,
Mr. Henry Jenner, and Mr. Henry Shrapnell, were
with him in a few minutes, and administered the
most judicious remedies. A messenger was sent for
me ; and I reached Berkeley about two in the afternoon. I found him in bed, lying in a complete state
of apoplexy. The right side was paralysed ; the
pupils of the eyes contracted to a point, and unaffected by strong light ; the breathing stertorous, with
a general insensibility to almost every external impression. Every effort was employed to arouse him
from this condition ; but the fatal character of the
malady became more and more apparent, and he expired about three o'clock in the morning.

And now, having brought this narrative to a conclusion, I would for a moment meditate on the character which, with affectionate regard, but most

heartfelt distrust of my own ability, I have endeavoured to delineate. Jenner stood in a position never before occupied by mortal man; having been the instrument in the hands of a gracious Providence, of influencing, in a most remarkable degree, the destinies of his species. He lived at a time when the whole of the civilised world was ravaged by a war of almost unequalled ferocity. Before he left the stage he had the supreme gratification of knowing, that his discovery had been the means of saving more millions of lives than had been sacrificed during the murderous conflict.

If we look at the origin of this discovery from its first dawning in his youthful mind at Sodbury, and trace it through its subsequent stages—his meditations at Berkeley—his suggestions to his great master, John Hunter—his conferences with his professional brethren in the country—his hopes and fears, as his inquiries and experiments encouraged or depressed his anticipations—and, at length, the triumphant conclusion of more than thirty years' reflection and study, by the successful vaccination of his first patient, Phipps; we shall find a train of preparation never exceeded in any scientific enterprise; and in some degree commensurate with the great results by which it has been followed.

In the space of a very few years, the fruit of this patient and persevering investigation was enjoyed in every quarter of the globe; and the rapidity of its dissemination attests alike the universality of the pestilence, and the virtue of the agent by which it was in many places subdued, mitigated, extirpated.

On the other side, let us remember his trials, his mortifications, the attempts to depreciate his discovery and to check its progress, together with the personal injuries which he endured from those who affected to do him honour, and we shall find many things to counterbalance the homage and gratitude which he derived from other sources. Under all these changes, he sustained the equanimity and consistency of his character; humble when lauded and eulogised, patient and forbearing when suffering wrong; and, if it be an assured sign of a worthy and generous spirit to be amended by distinction and renown, no man ever gave stronger proofs of possessing such a spirit.

Again, we have to view him in the character of a physician, exercising all the resources of a painful and anxious profession with extraordinary humanity, ability, and perseverance; cultivating his beautiful taste for natural history and all the poetry of life, in connexion with labours so arduous and important. While interpreting nature, he enjoyed a pleasure surpassed by none of his predecessors; but he did not rest there, and might have exclaimed with the great Linnæus, O QUAM CONTEMTA RES EST HOMO NISI SUPRA HUMANA SE EREXERIT!

As a husband, a father, a friend, a master, he may challenge comparison with any of his fellow-mortals. His domestic duties, as many traits in these volumes show, were invariably exercised with a degree of kindness, consideration, and delicacy, never exceeded. His attachment to his friends knew no variation or interruption, and even when his mind was almost overpowered by the pressure of his public engagements, he always found leisure to maintain and cherish his relations with them. He was not less mindful of his dependents and his neighbours in the humble walks of life: they were, indeed, his friends, and he treated them as such; and it is a graceful illustration of this principle, to see him building a cottage as a place of refuge for his first vaccinated patient, a few years before he died.

He was invariably courteous and generous to the stranger; "compassionate to the afflictions of all, shewing that his heart was like the noble tree, which is wounded itself when it gives the balm." He readily pardoned and remitted offences, proving that his mind was raised above injury, and could not be reached by the shafts of malignity. Finally, he laid down his life while continuing his efforts to do good to his fellow-creatures; grateful to God for the signal mercies which He had vouchsafed to man through him.

As soon as the melancholy event of his death became known in London, some of his friends were particularly anxious that he should have a public funeral. Sir Gilbert Blane felt that Westminster Abbey was the only fit place to receive his remains; and moved by the impulse of his own just and generous mind, he instantly wrote to the relatives of Dr. Jenner, to propose that they should concur in such an arrangement. Had those in power and authority viewed things in the same light, and ordered a public funeral at the public cost, Dr. Jenner's friends could not but have assented to the proposal.

As, however, no such authority was given by
government, and considerable expense must have
been incurred, it was deemed inconsistent with the
humility and modesty of Dr. Jenner's character to
seek for an ostentatious display of the pomp of woe,
however much he merited the sincere lamentations
of every well-constituted mind. It was therefore resolved that the arrangements which had been commenced before the proposal was received, should be
completed; and that his own Berkeley should hold
his remains; they were accordingly deposited in
a vault in the chancel, by the side of his beloved
partner, on Monday the 3rd of February, 1823. It
was intended that the funeral should be strictly
private; but all his personal friends who were within
a reasonable distance felt themselves constrained
to attend. The following list, I believe, compre-
hends a few of those who were present on that melancholy occasion. The Right Honourable Lord
Segrave, Colonel Kingscote, Colonel N. Kingscote,
T. Kingscote, esq. Dr. C. Parry, T. Creaser, esq.
Rev. Dr. Worthington, Rev. Dr. Davies, Rev. Mr.
Halifax, Rev. Mr. Ferryman, Rev. T. Pruen, Rev. G.
Jenner, T. Hicks, esq. H. Hicks, esq. R. Davies, esq.
John Hands, esq. H. Shrapnell, esq, Henry Jenner,
esq. John Fosbroke, esq. his only surviving son,
R. F. Jenner, esq. John Yeend Bedford, esq. and the
author of this work, &c. &c.

As death is supposed to shut the door of envy and
to open that of fame, it was hoped that the departure of so eminent a person would have been
commemorated by corresponding tokens of respect.

It was especially thought that all the leading members of his own profession would have eagerly seized such an opportunity of burying in oblivion every hostile feeling, and that they would have cordially and unanimously co-operated in rendering honour to his name, who had conferred such unexampled honour on their profession. The medical men in his own district were guided by this laudable and becoming spirit; but before they moved they were desirous of seeing some steps taken by the leading physicians and surgeons in the metropolis. As I had a considerable share in these transactions, it is due to myself to mention what actually occurred. I laid the matter before the late Dr. Baillie and other gentlemen, with a view of inducing them to attempt to call forth the exertions of professional men in London, in order that a conspicuous monument might be erected by them to the memory of Jenner. He having died in Gloucestershire, it was deemed advisable that the design should originate in that county. A meeting of medical men was accordingly held at Gloucester, on Saturday, the 22d of February, for the purpose of entering into a subscription, and making other arrangements for the erection of a Provincial monument in honour of the Author of Vaccination. The design was, that this object should be altogether accomplished by the contributions of professional gentlemen in different parts of the kingdom; and, under the expectation of a large number of contributors, a small sum was fixed upon as the amount to be subscribed. Our calculations proved erroneous. We did not

find that extensive co-operation either among the
learned bodies of the profession, or among individuals, which we anticipated. The only two public
bodies who contributed any thing, were the Colleges
of Physicians and Surgeons of Edinburgh; the former
having given £50, and the latter £10. After considerable difficulty, a sum of money was raised sufficient to enable us to give orders for a statue by
Sievier, which has been placed at the west end of
the nave of the cathedral of Gloucester.

On the 19th of March (1823), when the House of
Commons was in a Committee of Supply, and the
annual grant for the National Vaccine Establishment was voted, an attempt was made to obtain at
the same time a sum of money for erecting a monument to the memory of Dr. Jenner. Our excellent
county member, the late Sir B. William Guise, bart.
brought the subject forward, and it was met with
very considerable cordiality. He proposed that a
specific sum should be granted for the purpose.
This proposal was received with cries of " Move;"
but the Chairman of the Committee seems to have
quashed the business by stating, that a vote of
money could not be increased in the Committee.
Mr. Bright, the Member for Bristol, was anxious to
draw the attention of the Chancellor of the Exchequer to the merits of Dr. Jenner, and he hoped the
Honourable Baronet (Sir W. Guise) would move in
a more formal manner, and in a fuller house, for the
sum which he had mentioned.

I have reason to know that this design would have

been supported strenuously by many Members of Parliament: but it was never brought forward again. It ought, however, to have been mentioned, that the subject had been previously laid before the Government by Sir William Guise, on the 17th of February. " Seeing the Right Honourable the Secretary in his place, he requested to know whether there was any intention on the part of Administration to propose that a monument should be erected to the memory of the late Dr. Jenner. Sir William stated that he was induced to put this question, as having the honour to represent the county of which that eminent man was a native; and also because he thought this country was bound to shew that respect to Dr. Jenner's memory, which his valuable services, not only to this country, but to the whole world, by the discovery of vaccination, had so amply merited.

" The Chancellor of the Exchequer agreed entirely with the Honourable Baronet as to the merits of that eminent character, and the great blessings which he had conferred, not merely on his country, but on mankind; he confessed, however, that this subject had not previously occurred to him, and he was not prepared to say how far it might be expedient to propose such a vote as the Honourable Baronet had suggested.

" Mr. Secretary Peel said, that there was every disposition on the part of the Government to honour the memory of the distinguished individual in question; and it was only out of respect to the wishes and feelings of his relatives, that he had not been publicly interred in Westminster Abbey."

The concluding remarks of the Right Honourable Secretary would lead one to infer that the cost of the funeral was to have been paid by the public; and that Government had assented to so becoming an appropriation of a part of the wealth of the state. I have, however, already mentioned, that this was not the understanding of Dr. Jenner's family; and mainly on that account they declined the proposal. As, therefore, the public were not put to any charges on account of his funeral, there is still an additional reason for voting that a monument should be erected to his memory.

LETTERS

SUCH is the detail which I humbly offer respecting this eminent man. But I feel it impossible to do justice to his character, without permitting him to speak for himself, through the medium of a few of his letters out of many in my possession. The reader will therefore find a short series, bearing out, I trust, the accuracy of the preceding narrative. The first illustrate some important points in the early history of vaccination. Others, chiefly apply to the diffusion of the practice over the globe, and the management of the National Vaccine Establishment. Lastly, will be found several of his familiar epistles, which more peculiarly throw light upon his domestic habits and occupations.

I intend to add two or three of his metrical compositions which have fallen into my hands since the publication of the former volume. These,

Y 2

with some of his speculations on miscellaneous subjects, will heighten the colouring of the picture: more might have been added, but for the present, I am induced to withhold them, lest this work should expand to an inconvenient extent. As it is, I would hope, that " actions, both great and small, public and private, have been so blended together, as to secure that genuine, native, and lively representation which forms the peculiar excellence and use of Biography."

TO THE REV. JOHN CLINCH, TRINITY HARBOUR, NEWFOUNDLAND.

Cheltenham, July 15*th,* 1800.

MY DEAR FRIEND,

MY pursuit, thank God! is constantly making those advances which increase my fame, and will certainly add to the stock of human happiness, by eradicating one of the greatest of its miseries. Lest the threads sent you by George should not take effect, I have inclosed a bit more, newly impregnated with the cow-pock virus; use it like a small-pox thread; but small as it is, divide it into portions, that you may multiply your chance of infecting. Wet it before insertion, or rather moisten it.

My acquaintance with your Governor commenced from my having inoculated his infant daughter. I hope you have got my books on the subject. I am now just got to Cheltenham, having spent near six months in London. This business occupies, as you may suppose, much of my time and attention. A man of the name of Brown, a surgeon in London, has made a variety of efforts to write it down; but finding himself deserted by every medical man of respectability, he shot himself a few days ago. In every case of this sort, I have charity enough to suppose it is the maniac and not the man who draws the fatal trigger.

God preserve us, my dear friend, to see each other once
more!

My best remembrances to Mrs. Clinch.—Adieu,

Yours most affectionately,

E. JENNER.

To H. HICKS, ESQ. EASTINGTON.

1801.

MY DEAR FRIEND,

I think you are perfectly right in putting a stop to the
further insertion of the advertisement. I know not what
the people here will do, as they are quite beyond my control;
but most likely they will do nothing. However, on that
score make yourself easy, for so many awkward accounts
have by design and accident crept into the papers, respecting the intention of the subscribers, that I do assure you it
may have no bad effect (if well managed) should something
appear from some of the London people.

One thing I can answer for; the public speak highly of
the measure Lord B—— has brought forward, and it will
certainly aid, not injure, the main object. I am now fully
prepared to meet the House of Commons, and defy all the
scepticism that can be produced, to stand a moment before
my mass of facts.

A famous paper has been transmitted to me from Amiens.
The medical department of the Somme has sent a long letter to Lord Cornwallis, the purport of which is a compliment
to the British Nation, on its having given birth to the happy
discoverer of vaccine inoculation. This letter some of my
friends here entreat me to send to the newspapers, and it is
probable I may comply; but not as coming from me. Taylor has completely stopped Lord Holland's mouth by his
last despatch. I was at Holland House in a little time after
it was received.

A change in administration is certainly to take place, but

I don't think a very important one. All I *know* is, that
H——se and T——y are certainly to join the ministerial
phalanx as officers in some shape or other. Every petition
like mine goes, or seems to go, to the King, and comes from
his Majesty to the House. This, I believe, gives birth to the
form you see now and then in the papers. Darke, when at
Cheltenham, mentioned some strong cases to me of the preventive power of cow-pox. He can also favour me with cases
of those who have resisted variolous inoculation, because
they had undergone the cow-pox at some distant period of
their lives. Evidence of this kind I cannot obtain too abundantly, as it is at this point the public mind makes a pause,
from the early impression that was made of its proving a
temporary preventive only. This must be the form : first
state the evidence of the preventive powers of cow-pox, and
then add any comment you please upon the utility of the
discovery. You may compare the anxiety you felt on the
variolous inoculation in your family, with your feelings respecting the vaccine. Say nothing of Paul. It is time
enough to determine how the subscription money shall be
disposed of. A gold cup I should make choice of, in prefer-
ence to any thing else, if I may be allowed to name what it
shall be. Have you thought of an appropriate device, &c. ?
What think you of the *cow jumping over the moon?* Is it
not enough to make the animal jump for joy ?

My best regards. Yours truly,

EDW. JENNER.

To H. HICKS, ESQ.

Wednesday night, April 28th, 1802.

MY DEAR FRIEND,

You have doubtless seen Gardner before this time, and
heard many anecdotes of the Committee Room. Notwithstanding it is frequently dinned in my ears that the opponents to my claims will be strong, and that an adverse party is

actually mustering to take the field against me, yet I am not at all dismayed. The Admiral tells me I have nothing to fear. The Report will probably be made this week; after this, the evidence will be printed and lie on the table a day or two before the final discussion. I sometimes wish this business had never been brought forward. It makes me feel indignant to reflect that one, who has through a most painful and laborious investigation, brought to light a subject that will add to the happiness of every human being in the world, should appear among his countrymen as a supplicant for the means of obtaining a few comforts for himself and family. Upon my word I can hardly stand it, nor should I have stood it so long had it not been for Hobhouse. I told him one day in the Committee Room, (feeling chagrined at the treatment I there experienced) that I should be glad to put an end to the matter, and withdraw my Petition. His reply was, that such a proceeding was impracticable; it must go on. There is a fundamental error, in my opinion, in the conduct of the Committee. Having been put in possession of the laws of vaccination by so great a number of the first medical men in the world, namely, that when properly conducted it never fails, and when improperly, that it will fail; they should not have listened to every blockhead who chose to send up a supposed case of its imperfection: but this is the plan pursued, and if they do not give it up, they may sit till the end of their lives; for the inoculator of the cow-pox, like the small-pox inoculator, will go on for ever committing blunders. Within this fortnight an apothecary here, who attended a family where I was inoculating, attempted to take virus after the pustule was nearly converted into a scab. How unjust it is to make me answerable for all the ignorance and carelessness of others. You may depend upon my vigilance and activity. To-day I inoculated a son of Lord Holland: this will give me frequent opportunities, I hope, of seeing some of those great characters who will do more than support my cause in the House of Commons with a simple "Yes." Mr. Grey has promised his support, and Mr. Fox, I doubt not, will be amongst my special pleaders. Do you never intend visiting the metropolis? I cannot blame you for keeping out of it. Most heartily do we all pant for the country air. We are all looking sadly. I have taken a box for Mrs. J. and the children at Bayswater. My knocker has been tied up for several days, on account of poor Catherine. Thank God, to-morrow it will be unmuffled, as I think her out of danger. She has had a most violent fever, but not typhus. She had symptoms that seemed to indicate water in the head. Luckily this was not the case. I was thrown into extreme alarm, from which I thankfully acknowledge myself completely liberated. To-morrow night we are to be all in a blaze here. The preparations going forward in every street are unexampled. I have just seen one in which Billy P. cuts a conspicuous figure, with a few lines descriptive of the scene from the pen of P. P. The papers will no doubt give it to you. My kindest respects and Mrs. Jenner's to Mrs. Hicks.

Yours most truly,
E. Jenner.

To Henry Hicks, Esq. Eastington.

10th May, 1802.

Dear Harry,

I have spoken to Dr. Heberden, and was much surprised to hear him say the Gloucestershire estates were not his but his brother's. This brother I will find out without delay, and sound him relative to the business.

You will forthwith receive a drawing of the emblematics for the cup: I have not seen them myself. The idea of the cow trampling, as if by accident, and then crushing to death the monster small-pox, I don't much like. This, I know, was the first design; perhaps you will find it altered, and the cow made to assume less of a pacific character. Your inscription is, I think, extremely appropriate, and strikes me as finishing well without the optional sentence; and yet if something to this effect could be well brought in, it might render the history more complete: however, after all, it may be best for the reader to have a trifle left on which to exercise his own imagination, and the word " small-pox " won't associate well with good drink.

With what sort of stuff Parliament will allow me to fill it, is not yet determined. It ought to overflow with nectar when presented to your lip. The Report was made to the House on Friday, and will be presented in two or three days for the inspection of the Members. Admiral Berkeley has quitted the chair for this fortnight through a gout-fit. It has been filled by Mr. Bankes.

I understand an opposition is to be made in the House on this ground—that it has been stated to the Committee, that I might have filled my pockets by the practice of vaccine inoculation, had I not made every body acquainted with it. How absurd! While I thus had been employed in filling my own purse, should I not have indirectly been filling the churchyard with those slain by the small-pox? This is really to be the plea for non-remuneration, as shewing there existed no necessity for my petition to Parliament. We want not the strength of a Fox or a Grey to combat this argument.

As usual I must apologise for omissions.

Yours truly,

E. JENNER.

Pray see the European Magazine for the present month, where you will find that the Indians have adopted vaccine inoculation: this should be conveyed to the Glo'ster paper.

The drawing of the cup is just sent by the Stroud coach by Lady B. You will hear from her by to-morrow's post.

To Richard Dunning, Esq. Plymouth Dock.
London, Monday Evening, May 17, 1802.

I must confess, my dear friend, that you have too often had occasion to rebuke me; but to have answered the letter you allude to fully and completely was a task so arduous, that truly (although often in contemplation) I put it off with the hope of finding that leisure for its accomplishment which has never yet arrived.

" Ye gentlemen of Plymouth,"

—— I was going to parody a popular song, but the attempt is too daring—*verbum sat*. I once thought myself the most perfect animal mimosa existing; but you seem to possess a superior claim. You will believe me sincere in my assertion; but don't be affronted when I tell you that Trotter (whom I had frequently the pleasure of seeing during his residence in town) and myself even laughed at your timidity.

If I have excited a frown, let me smooth your countenance by saying, that while I indulged this passion, I admired your sensibility. Your excellent letters deserve a better answer than this; but the worries of the House of Commons, and the labours I have to go through out of the House, put it out of my power to shew my friends even common civility.

I shall not conclude without telling you how affairs stand. The Committee have made their Report to the House, which is ordered to be printed. After it has lain upon the table a few days, it will then be taken into consideration, and I shall receive reimbursement, remuneration, or nothing at all; just as the honourable gentlemen may determine.

If the latter, I have to thank my stars that I have a little farm or two in Gloucestershire, where I shall at once repair, quit *doctoring*, and turn ploughman.

How happy shall I always be to give you a rasher of bacon from a *flitch* of my own fatting, and a potatoe that has vegetated under my own eye. I sing a little sometimes, and am practising Shakspeare's old song in readiness.

" Blow, blow, thou winter's wind."

I shall, if *possible,* send you one of the printed Reports. A line from you to Sir W. Elford would procure it at once. You will be amused with the intelligence from the West— the Drews, the Bragges, &c. &c. The conduct of P——— was bad beyond all description.

The evidence in the Report is so compressed, you will not be able to form by any means an idea of it. The letters from the West are printed at large.

I am much gratified at the good sense manifested by the Cherokee Indians. Who would have thought that vaccination would already have found its way into the wilds of America ? Pray look at the European Magazine for the present month. Be assured that I am yours most sincerely,

E. JENNER.

P.S. I shall expect the gratification of a letter from you very soon, especially as in your last you explained nothing relative to the cases which weakened your confidence in vaccine inoculation.

I laid before the Committee a large bundle of letters on the subject of cow-pox, corroborating the fact I have alleged, and expressed a wish that some of them might be printed in the Report.

I even limited my number (finding them shy of printing) to three, and among these named yours ; but, alas ! they indulge me with one only, and that comes from a Mr. Kelson.

By the evidence of a Gloucestershire gentleman, Mr. Gardner, you will see that I spoke freely of my scheme for eradicating the small-pox previously even to the capricious inoculations made by the gentlemen in the *West*.

To R. DUNNING, ESQ.

MY DEAR SIR,   1802.

Our last letters crossed each other on the road, according
to custom. Your letter of April the 22d reached me at a
time when my head was brimful of the bustles of the Committee, and was not, I think, sufficiently, at least properly,
noticed in any subsequent letter of mine. What I allude to
is your account of the inoculation of Mr. Courtney, Mr.
Yonge, and the staggering cases of yourself and Mr. Lisle.
Add to this, the case of the marine at Portsmouth. Now,
my good friend, my mind having long since obtained what
security it is capable of possessing, I request of you to tell
me what time and enquiry have developed respecting these
Plymouth cases. That of the marine at Portsmouth was
clearly made out to have been imperfect. The people at
this sea-port set up a kind of malignant shout (see the Letters of Hope in the Report of the Committee) at finding this
case of supposed failure. They disliked vaccination, because Plymouth adopted it; " tanta est discordia fratrum."
You mention the name of my valued old friend Col. Tench;
pray can you find out where a letter would reach that gentleman? Now to present occurrences. Imprimis, accept
my thanks for your benevolent intentions in seconding the
views of the society established, or rather about to be, for
forming a something (I scarcely yet know what) to perpetuate my name, and to attach to it that, without which
scarcely any existing name has weight or respectability. I
immediately despatched your note to Dr. Lettsom, who is
one of the great promoters of the design. I need not, I
presume, tell you that no idea of this sort ever came into my
own head. Let me entreat you not to suffer that warmth of
heart which you possess, and that affection (your deeds
allow me to use the expression) which you shew for me, lead
you too far on this occasion. The portrait by Northcote,
I conceive to be finished by this time; but as it is deemed

not only one of his best paintings, but what on the present
occasion is more to the purpose, a good likeness, a print will
be required before it reaches its destined home. Pearson,
you say, has exhibited himself with sufficient correctness.
Indeed, I think so too. Had he not given the picture those
brilliant touches which you have doubtless seen since the
publication of the Medical Journal for the month of July, I
trust that most people would have thought his pencil had
gone far enough; but what, my dear sir, will they say, now
they have seen his last performance? I long to hear your
sentiments upon it, and to know what sensation it creates
among my friends in your part of Devon. The paper by
Dr. Crawford of Bath was, I think, pretty much *ad rem*,
and will serve as a groundwork for any one who may
choose to reply to the pamphlet. As for myself, I do not
intend to notice it. To what purpose is it to contend with a
man whose arrogance and impudence will ever keep at an
immeasurable distance plain truth, and who will not listen
even to the dictates of reason? There is among the mass of
misrepresentations one which is capable of doing vast mischief, and that is, allowing the vaccine virus to be used in
the far advanced stage of the pustule. Woodville's backing
this opinion is quite astonishing. If I should be induced to
reply to anything, it would be this point only. To what reverend gentleman I am indebted for congratulations, and who
speaks so highly of Dr. Booker's excellent sermon, I know
not; this you keep a secret. The fact is, neither myself nor
any one about me can decypher the characters you have
employed to point out his name. However, present, I pray
you, my thanks to my unknown friend for his kindness, and
I hope ere now he has well vaccinated the ears of his congregation. The example of Dr. Booker has been followed
by many respectable clergymen, both in town and country,
and with good effect. Be assured I wish much to print the
evidence you allude to. It certainly ought to come forth,
and for the reason you assign.

4

Mr. Bankes, who drew up the Report, was no friend either to me or my cause, or he would have listened to my solicitations, and inserted not only the certificates you mention, but your letter also. Let any one read the Report of Dr. Smith, and compare it with mine; then let them judge who had indulgences and who had none. The indisposition of my chairman, Admiral Berkeley, was a most unfortunate event. The whole merit Mr. Bankes allowed me on the score of discovery in vaccination (considering it abstractedly) was that of inoculating from one human being to another. On this subject I remonstrated, but it was all in vain. Cannot you contrive to get your papers into the Journal? Surely you might command my assistance whenever you please; they would gain admittance with the most perfect propriety in reply to Pearson's audacious assertion, and produce good effects in a variety of ways.

My health and prosperity are not more frequently drank under your roof than that of you and yours is wished for under mine.

Adieu, my dear Sir,

Yours ever very faithfully,

Cheltenham, Sept. 24, 1802.   EDWARD JENNER.

I shall remain here about six weeks longer, then go to Berkeley, and stay about two months before I return to my drudgery in town. Do come and see me at Berkeley, if you can, at *Vaccina Cottage.*

TO R. DUNNING, ESQ.

MY DEAR SIR,   *Berkeley, April* 2, 1804.

The sight of your folio sheet was extremely agreeable to me. I really began to be seriously alarmed about you; but all is well now. You have passed upon yourself too heavy a censure. It is my misfortune to be a great procrastinator as well as yours; and surely it would be uncharitable in me not to pardon that in another with which unfortunately I find myself so heavily laden. Your letter throughout

glows with that ardent philanthropic flame which has ever
shone so conspicuously in all your conduct towards me from
the commencement of our acquaintance.

I believe we have had no correspondence since your *Spanish paper* appeared in the Medical and Philosophical Journal. To be plain with you, and use the familiarity of a
friend, I did not like it. The paper is not now before me;
but if I recollect right, it went only to prove that goats are
subject to spontaneous pustules upon their nipples; that
the matter of these pustules was inserted into the arms of
human subjects; and that it produced local effects. Is
there any quadruped that is not subject to diseased nipples?
Even the human animal, we know from sad experience, is
not exempted. The cow, like other animals, is subject to a
spontaneous pock upon its teats, the fluid of which, when
brought in contact with the denuded living fibre, is capable
of exciting disease; but I positively assert, this is not one
grand preventive. When you hear again from Madrid, do
not fail to tell me what the Spaniards say about it. I have
already anticipated. I am happy to find that Mr. Williams
has pleased you so much in copying Northcote's portrait of
me. Some of my partial friends say, that the painter gave
the countenance a dash of acid which does not belong to it,
but really the print by Saye appears to be a good likeness.

Dr. Hope, with all his prettiness about him, will continue
with me an object of distrust until he *studies* vaccination, and
having seen the light, tells the world he was in darkness
when he sent that horrid murderous letter to the Committee
of the House of Commons, first, indeed, to the Board of
Admiralty. I call it murderous, because by retarding the
progress of vaccination, it has sent many poor victims to
the grave who now might have been actively employed in
the defence of their country; and more will certainly follow
unless he publicly makes known his errors. Surely the finger of Providence has pointed out to him the spot he now

resides in, a spot which the votaries of Vaccina must ever contemplate with delight. He certainly made a most violent effort to ruin me and the cause in which I was engaged; but weak and pernicious as his conduct has been, I will forgive the whole transaction if he will behave like a man, and act as he ought to do. A man, in my judgment, never appears more wise or more amiable than when renouncing false opinions. I think you had better once more write to my Lord King, as it is probable his work will go through another edition, when he may have an opportunity of correcting his errors. I should be sorry if any bookseller had advertised any thing new from me, or any new edition of my work, though I long to republish it. In a former letter I believe I mentioned to you the effect that herpes often produces upon the arm; it is of so much moment that I have thoughts of giving a short paper on the subject in the Medical Journal. The useful terms "vaccination and to vaccinate" are undoubtedly yours, and as such I pronounced them at a meeting of the Royal Jennerian Society, when an M. D. present mentioned them as imported from the Continent.

In arranging a bundle of papers which had been huddled up together a long time ago in London, I met with my answer to a letter from the Medical Society of Plymouth. Whether it is a copy of a note already sent, or whether through hurry I might never have sent it at all, I cannot now recollect.

I must therefore beg your assistance. The idea of slighting, in the smallest degree, a body of gentlemen, who, on all occasions, have stood forth so strenuously in my support, would vex me very much. Will you then have the kindness to make inquiry of your secretary, Mr. Woolcombe, and to present the inclosed note (if it appears that I had not written upon the same subject before) with an explanation and many apologies.

I will not suppose my friend Little can have any other

impression made upon his mind by the occurrence of such events as you mention, than such as an enlightened physiologist would receive.

Pray thank Mrs. Dunning and your family for the kind compliment they daily pay me, and be assured of the friendship and esteem of

EDWARD JENNER.

Do not forget to tell your brother vaccinists to look sharp to tinea capitis, sore eye-lids, ears, &c. I have commonly remarked, that the impediment to the correct progress of vaccination arises more frequently when the disease is recent than when it has continued long.

If any of the *birth-day odes* and verses were committed to the press, you would indulge me in sending copies.

Do not fail to write soon. I want to know your further sentiments of the goat-pox.

April 5.

Finding my letters over weight for a frank, I make a packet, and send it by the coach. The little pamphlet of Warren has scarcely ever failed to make converts whenever I have sent it into a poor prejudiced family.

I have just received the Portsmouth paper of the 2nd of April, sent to me, I suppose, by the printer. It contains, in *large letters*, the following sensible paragraph : " Reports of some cases of small-pox after vaccine inoculation were read at a very full meeting of the Medical Society of Portsmouth on Thursday last the 29th instant, which we are informed will be sent to the press, and published in a few days." Is Dr. Hope returned to his old post ? What a set of blockheads ! How will our continental neighbours laugh at us !

A letter from Milan lately informed me, that Dr. Sacco has, with his own hands, vaccinated upwards of 40,000 without a single instance of the small-pox occurring afterwards ; and why ? He understood what he was about.

R. DUNNING, ESQ.

Berkeley, July 22, 1804.

MY DEAR SIR,

Your letter found me in the midst of the worries attendant on packing up. You are a stranger, I presume, to the removal of a large family from place to place; if you are not, you must know how to commiserate those who are. Accept this as an apology for delay in writing. Even gentle hints of mine, I perceive, act as mandates upon you. It affords me pleasure to see with what alacrity my troops fly to arms, and rally around me at the approach of an enemy.

> " Horse and foot
>   All fly to it."—*Old Ballad.*

Seamen and landmen are all ready. The Portsea corsair, already pretty well peppered by the *King,* the *Edinburgh,* and the rest of the squadron, must strike at the first broadside of the *Dunning man-of-war.* But to quit figurative language and descend to humbler tones, allow me to say, I look forward with great satisfaction to the result of your present undertaking. Plymouth must be able to furnish volumes of evidence, were they wanted, to refute the absurd arguments of Mr. Goldson. All his reasoning is erroneous. It must be so; for how could he reason upon a subject of which it is plain, from his own words, he has scarcely any knowledge? A man who takes up the pen in such a cause, should be intimately acquainted with the laws and agencies, not only of the vaccine virus in the constitution, but with those of variolous also. Now 'tis clear he knows neither. What a book is his! What an advertisement did he send to the public papers! An advertisement that will outstrip its keenest pursuers, and strike terror as it goes into the bosoms of thousands. Our newspapers are spread over the face of the whole earth. The trials I have lately instituted here, assisted by my nephews, I can assure you, have been severe

ones; but, thank heaven! they have been decisive, and without any other aid must completely overthrow the arguments
of Mr. Goldson. All the subjects that I could collect, who
were vaccinated at the commencement of my practice here,
men, women, and children, have been lately exposed to the
small-pox, in a state as highly contagious as possible; they
were taken into a room, and went to the bedside of a woman,
covered from head to foot with pustules. All have escaped
unhurt *except at the sight of the ghastly object.* A great
number of these had been inoculated six years ago. Phipps,
too, the boy on whom I made my first trial more than eight
years ago, has again been put to the trial with impunity.
Had Mr. Goldson sought an interview with me, or even
written to me on the subject, I am confident his book would
never have seen the light. Perhaps it may be all for the
best. Had vaccination wanted firmer support than it has
already, it would have obtained it from the very efforts
made use of for its destruction. I will just remark that the
fairest of all tests is exposure to variolous contagion; this is
the natural test, inoculation is not. Who does not know
(all medical men ought to know) that the insertion of the
variolous poison into the skin of an irritable person will
sometimes produce great inflammation, disturbance of the
system, and even eruptions?

Adieu, my dear Sir. I write, as you must observe, in
haste.

Yours truly,
E. JENNER.

Just setting off with my family to Cheltenham.

P. S. I am sorry to say I cannot send you advertisements
to the cover of the Medical Journal. The review of G.'s
book will tell you I have no interest there.

R. DUNNING, ESQ. PLYMOUTH.

Cheltenham, July 25, 1804.

MY DEAR SIR,

About the middle of the last year, Dr. Yeates, a physician of eminence at Bedford, published a pamphlet explaining the cause of some vaccine blunders that had been committed in his neighbourhood. It concludes with a few observations of mine upon the subject. Conceiving you may never have seen this little tract, I have sent it to you; for, like the tracts of *some* folks in days that are past, it was never well advertised, and consequently but little known.

I am quite angry with myself for not noticing in my last the Plymouth Bard. You call him a Printer's Devil. If spirits of this description can send forth such things, what am I not to expect from Plymouth Gods.

Yours truly,

E. JENNER.

RICHARD DUNNING, ESQ. DOCK, PLYMOUTH.

Cheltenham, October 25, 1804.

MY DEAR SIR,

Before I say anything of your second letter, allow me to notice your first. When I tell you that I am at this time at least two hundred letters in arrears to my correspondents, which, as you may suppose, multiply in every part of the earth to a great extent, you will at once forgive my not writing sooner to a friend with whom I could take a liberty. There is not a country in the globe where I do not owe a letter, and yet all my leisure time is occupied with pen, ink, and paper. But you must be informed that my leisure hours are very few; for the company resorting to this fashionable watering place increase every year in a most rapid manner, and, consequently, my medical engagements; insomuch, that I have it in contemplation to quit it. Should I be compelled to do this, what a hardship must I endure!

Shall I not be the first man in our profession who quitted his post through excess of business? Vaccination calls imperiously for my attention, and to that I am determined all my other worldly concerns shall yield. But while I am fighting the enemy of mankind, it will be vexatious to see my aides-de-camp turn shy. Among the foremost in the field, I have always ranked *Richard Dunning.* No one has been more obedient to the commands of his general, or wielded the sword against the foe with greater force and dexterity. But shall I live to see my friend dismayed at the mere shadow of fortune on the side of the enemy;—will he who has led such hosts into the field and found them invulnerable, start if, in the continuation of the combat, he should see a man fall? Enough of metaphor. The moral of all this is, that I see you are growing timid; the timidity so conspicuous towards the close of your pamphlet, and that which is so manifest in your letter of this evening, it would be wrong in me not to say I was sorry to observe. More convincing or stronger facts the public could never wish for than your pamphlet exhibits. Had I been at your elbow, I should have certainly pulled back your pen when you began reasoning upon them. The result of your experiments authorised you to speak in tones the most exulting and triumphant; but most unfortunately, you almost give up the field to the anti-vaccinists by speaking of new and better arrangements, IF *variolous inoculation should supersede the vaccine!* Now, my good and valued friend, don't for a moment think that I am out of temper with you, or mean to speak harshly. On the contrary, I attributed this oversight (such I must call it) to the dreadful calamity that befel your family. Your mind, I know, must have been oppressed, and you were bringing your work to a conclusion under pressures scarcely bearable. To those who made remarks upon what appeared so extraordinary, I communicated the circumstance which seemed to me to account for it. The 115th page of your work, is that which has occasioned the general surprise. The further I go on with vaccination, the more I am convinced that the great and grand impediment to the correct action of the virus on the constitution, is the co-existence of herpes. I expected that my paper on this subject in the Medical Journal for August would have attracted more attention. Since my writing it, I have detected a case of small-pox after small-pox inoculation, where the cause of failure was evidently an herpetic affection of the scalp. Are such cases as these—are such as Mr. Embling, so circumstantially described in your pamphlet—are Mr. Trye's, lately communicated in the Star—are Mr. Kite's of Gravesend, and a thousand others, to go unnoticed by the public, while failures in vaccination (a science far more difficult to understand than variolation) are to make impressions so deep as even to stagger the faith of those who are well informed upon the subject? Is common sense to be attached to one side of the question only, and to have nothing to do with the other? "This case, connected with those in London at Fullwoods Rents, I grieve to say, appear extremely ugly."— *Dunning.*

Is it possible their ugliness can affright you? What phantoms must they appear, if you will but look back and consider the period when those children were inoculated. Woodville at that time, and his coadjutor Waschell, knew nothing of the cow-pox; this is clearly evinced by Woodville's first pamphlet, where he gives three hundred cases of small-pox, and calls them cow-pox. Surely his early inoculations are not to be regarded; and does he not at this hour, in conjunction with a person whose dirty name shall not daub my paper, sanction the taking of virus from the pustule at any of its stages? What are we to expect while such things as these are going forward? Inclosed is the letter you requested me to return; it is impossible for me to go into particulars on such cases. I can only go into general reasoning. My experience justifies me in saying that which I have said fifty times before, " If the vaccine pustule goes through its stages correctly, the patient is secure from the small-pox; if not, security cannot be answered for." There certainly is sometimes a nicety in discrimination, and it was this which in my early instructions occasioned me to say, " When a deviation arises in the character of the vaccine pustule, of whatever kind it may be, common prudence points out the necessity of re-inoculation." Cases may possibly occur, where even you or I may (from the interposition of those events which medical men are always subject to) not have it in our power to catch opportunities of passing our judgment upon a pustule during those stages, whether it is or is not correctly defined. With respect to the doctrine of Mr. Moyle, I must candidly say, my experiments do not justify me in subscribing to them. Be of good cheer, my friend. Those who are so presumptuous as to expect perfection in man will be grievously disappointed. His works are and ever will be defective. Let people, if they choose it, spurn the great gift that heaven has bestowed, and turn again to variolation. What will they get by it? Let them consult pages 67 and 68 of your decisive work on this subject, and they will know. Let them peruse the following extract from a letter which I have within these few days received from a medical gentleman of great respectability in this county. " A poor family belonging to Sudeley parish, consisting of a man, his wife, and five children, were vaccinated four or five years ago, except the eldest daughter, who had been before inoculated for the small-pox by an eminent practitioner, and pronounced secure. This summer she caught the small-pox when working among the rags at the paper mills, and had a very numerous and confluent eruption. The rest of the family had no fears, and have all escaped, though fully exposed to the infection." Now had this case been reversed, what a precious morsel it would

have been for an anti-vaccinist. Adieu, my dear friend, and
be assured of the unalterable regard of

Yours,

EDW. JENNER.

P. S. I must not forget to tell you that I have sent your
pamphlet to the National Institute of France. They had
received Goldson's book, which I perceive was disseminated
with uncommon industry. I was not a little hurt at Mr.
Embling's taking no notice of a letter I deemed of some
importance, and which I wrote to him immediately on seeing
his observations. One word more on herpes. Seeing how
frequently the vaccine disease becomes entangled with it,
my thoughts have lately been pretty much bent upon it, and
I now see that the herpetic fluid is one of those morbid
poisons, which the human body is capable of generating, and
when generated, that it may be perpetuated by contact.
Children who feed on trash at this season of the year are apt
to get distended bellies, and on them it often appears about
the lips. This is the most familiar example I know. A
single vesicle is capable of deranging the action of the vaccine pustule. Subdue it, and all goes on correctly.

To R. DUNNING, ESQ.

Cheltenham, Nov. 2nd, 1804.

MY DEAR SIR,
Inclosed is Mr. Moyle's letter. Conceiving it might be a
gratification to you to see how systematically they manage
vaccine affairs in India, I have sent you a copy of a paper
just transmitted to me from the India House. Would to
Heaven we could boast of such arrangements here!

Yours truly,

Don't forget Mr. Embling.

With a view of extending the practice of vaccine inoculation throughout the East India Company's territories in India, the Governor General in Council of Bengal has appointed a superintendant general of vaccine inoculation at the presidency, and established subordinate superintendents at several of the interior stations of the country; viz. at Dacca, Moorshedabad, Patna, Benares, Allahabad, Cawnpore, and Farruckabad. These superintendents are the surgeons of the stations, and are to act under the orders of the superintendent general at the presidency in whatever regards vaccine inoculation. The civil surgeons also at the several judicial and revenue stations are to co-operate with these superintendents for the purpose of forwarding the general object. Vaccine inoculation has also been introduced with success into Prince of Wales Island, and it is intended to extend the practice to Malacca, and other places to the eastward, and a confident expectation is entertained that the benefits of this valuable discovery will be diffused throughout Asia. It is even in contemplation to extend it to China; but as the suspicious disposition of the Chinese might possibly ascribe any attempt to introduce this novel practice to sinister motives, it has been postponed until the opinion of the Company's servants there can be obtained.

Fort William, 13th Jan. 1804.

Nov. 12th.

I shall expect very soon to hear from you. Pray give me your sentiments with as much freedom as I gave you mine in my last letter respecting vaccination, &c. &c.

No frank to be had to-day; but I do not think you will regard the tax it imposes on you for the inclosed extracts from our provincial newspaper. They would not cut bad figures in that of Devon. The letter in the Times, in reply to Moseley, (copied into the Medical Journal under the signature of a Looker-on) should be generally read, as it prepares the mind for the consequences of incorrect vaccination. Addington's letter in the Morning Chronicle of the 5th of

November is firm, manly, and decisive. Did it catch your eye? From the partial evil of now and then an imperfect case, what an immense mass of good springs up.

To Richard Dunning, Esq. Plymouth.

Cheltenham, 15th Nov. 1804.

My dear Sir,

The old occurrence of our letters crossing on the road has, I see, again taken place. If my writing frequently to you will afford you the least gratification, I shall not be slack in my correspondence.

There is no one more entitled to my attention, and among all the vaccinists who have enlisted under my banner, there is no one who has a greater claim to my regard. There was no expression in my letter, I hope, which would bear the construction you seem to put upon it. You were rallied a little on your timidity respecting the *ugly* cases in town and country—on your glancing at a better regulation for the management of the small-pox, *if* we are obliged to turn to it again—on your fear of reviews—and of a little shrinking, even, from the man whom you are opposing; but all was done in perfectly good humour, and now you will allow me triumphantly to exclaim, "Richard's himself again!"

Dr. Borlase is an old acquaintance of mine. I have too good an opinion of him to suppose for a moment he could invent such an idle tale as that which has been told you by Mr. Moyle. The whole is an abominable fabrication. Indeed such an assertion from me would have flown in the face of those facts which my experience on this subject has rendered decisive.

Is it possible Goldson can appear again in print on the vaccine subject? Your communication is the first that has been made to me respecting it. He had better be silent unless he addresses the public in the humble, yet honourable, strains of recantation; for with all the supposed imperfections on the head of *Vaccina*, there are ten times as many on *Variola*.

Pray indulge me with a line or two very speedily, to put an end to a little perplexity. You tell me that you know small-pox will sometimes follow cow-pox, and nevertheless assert that a case of this sort, which has happened under your immediate observation, places vaccination on higher ground than it has yet stood on.

Do pray explain, as soon as you can, your meaning.

Yours, dear Sir,
very truly,
E. JENNER.

In the Morning Chronicle of Monday, the 12th instant, is another letter of Addington's.

I am pleased at seeing the friends of the vaccine cause shewing themselves in the newspapers. These meet every eye, while the Journal meets that of medical men only, and has proved the tomb of many an impressive paper.

To R. DUNNING, ESQ.

Berkeley, Feb. 10*th,* 1805.

MY DEAR FRIEND,

Your *little* pamphlet contains many great and useful observations. I will now refer you to a few notes I made in perusing it. The book itself should have been printed in the more general shape and form of pamphlets. Page 16, concluding sentence of the first paragraph, pithy, and containing a complete reply to the anti-vaccinists, who may urge objections from a few solitary cases of small-pox after cow-pox, or who might bring them forward if they were ten times as numerous. 100,000 cases of vaccination, by far too few to calculate upon. Half that number I can reckon from extra professional inoculations, 20,000 of which are from my fair disciples; and, to their credit be it spoken, I have not heard of one sinister event among this class of inoculations.

And why? They implicitly obey vaccine laws. Page 12, good reasoning on the subject of population. I have often urged the following argument when too numerous a population has been thrown in my teeth, as one of the ill effects likely to attend vaccination. Who would have thought, a century ago, that Providence had in store for us that nutritious and excellent vegetable the potatoe—that ready-made loaf, as it were, and which is prepared in higher perfection in the garden of the cottager than in the highly manured soil of the man of opulence? It is but reasonable to suppose that the same Omniscient Being that showered down this blessing on our heads, has similar stores in reserve, of which we of course can form no kind of conjecture. One thing we do know, that the science of agriculture is as yet only in its bud.

Your manner of speaking of Goldson increases his arrogance. He obstinately holds the veil before his eyes, and will not behold the vaccine light. I am about to make a stronger pull at this veil than has been done yet. I have sent him an invitation to visit me at Berkeley, or to appoint a deputation from the Medical Society at Portsmouth; I have gone further (perhaps too far), I have almost pledged my word that his conversion will be the consequence of the interview. The fact is, he is totally ignorant of that wise discriminating power, without which no man can be a perfect vaccinist: and it is my wish to impart it to him. One might as well contend with a blind man on the nature of the prism, as with a person in this situation, and entertain a hope of being successful; but to proceed.—In another edition, pray take in Kite's cases of small-pox after small-pox inoculation. They are the more forcible as they were published antecedent to the vaccine practice. Page 38. Are you sure the pustule was variolated? Page 41. I do not see the necessity for your parenthesis. Perhaps my feelings are too acute, but I do not like to see my darling child whipped even with a feather. In your postscript, why not ask for cases of small-pox after small-pox inoculation, as well as cases of small-pox after vaccination?

Have you seen Moseley's infamous pamphlet? You ask for fatal cases of the vaccine. This gentleman, in one single paragraph, furnishes you with some of the most terrible deaths that ever were heard of from this cause. One would suppose he was speaking of the small-pox, as he tells us the children did not lose their torments even in the article of death! Luckily, he takes away every thing like truth that can attach to this history by omitting every kind of reference. What punishment does a man of this description merit? Accept my best thanks for your letters. I am happy in seeing the unanimity that prevailed at the festive board. I have read King's observations in the Journal, but in the most cursory way only: the book was then taken away from me, and I cannot now refer to it,—are you certain you clearly understand him? In the cicatrix of those children on whose arms, through the intervention of herpes, the pustule has proceeded irregularly, I find in general a singular deviation, which it will be difficult to describe by words, and I draw most wretchedly. Instead of the flat, correct indentation, the cicatrix exhibits a perceptible elevation of a conical shape, though very slightly so. I have a fine specimen, in a child lately inoculated, with recent tinea capitis, and shall endeavour to take a cast in wax or Paris plaster. Indeed a series of pustules might be done in this way, and afterwards coloured. Is it possible any one can be so absurd as to argue on the impossibility of small-pox after the vaccine? I trust my friend King is not one of them. I hope to spend a month in town this spring, which time I shall devote entirely to the service of Vaccina. May, I believe, will be my month. I shall order a score of your pamphlets from Murray's.—You make me smile at the mention of " Berkeley booksellers." I hope one day or another, you will come and see *our village*. It has the name of a town, but in size it is a mere village. I shall send your duodecimo to my friend

Trye, who, I am sure, will peruse it with pleasure. Trye
is not only an eminent surgeon, but a very excellent man.

Yours, dear Sir,
With great regard,
EDW. JENNER.

P. S. I hope you will soon get a pamphlet sent from
Bengal, and republished here—" Report on the progress of
Vaccine Inoculation in Bengal, by John Shoolbred." Have
you seen by the papers what an ally the anti-vaccinists have
won over to their interest? Dr. Brodum has at length
joined the forces of the enemy.

TO THE REV. JOHN CLINCH, TRINITY, NEWFOUNDLAND.

Berkeley, August 16*th*, 1805.

MY DEAR FRIEND,

G. T. has just informed me that he has lately received a
letter from you, and that had it not been for a "fierce catarrh" that harassed you for two months in spring, every
thing would have gone on well with you. If this malady
should dare to molest you next year, retreat, seek the milder
shores of Old England, and leave the land of snows and ice
to the bears, for whom nature made it. Full four months
ago, I began a letter to you; but intending to make it a long
one, after writing two pages I mislaid it, and never have been
again able to recover it.

I went to London in May, and stayed till August the 3rd.
I am now packing up, (oh! how I wish I had you at my elbow,
you are such an excellent hand at the arrangement of a box,)
and setting off with my family tomorrow for Cheltenham, to
stay three or four months, then again to London. Never
aim, my friend, at being a public character, if you love
domestic peace. But I will not repine.—Nay I do not repine, but cheerfully submit, as I look upon myself as the instrument in the hands of that power which never errs, of

doing incalculable good to my fellow creatures. You would do me an essential kindness in acquainting me with the state of vaccination in your island, as I shall appear again before the House of Commons next session, and I am collecting all the information I can from foreign parts. Write to me not as if your letter was to be shewn to the House of Commons, and detail the real state of facts relative to the benefits derived from the new practice. Remember me kindly to Mrs. Clinch, and my old friend Edward, who I ardently hope is becoming useful to you; and believe me, dear Clinch, ever truly, and sincerely yours,

EDW. JENNER.

P. S. George will write to you more fully.

Many thanks for your kind attention in sending me such beauteous stores of fish and berries.

Do you recollect any cases of persons catching the small-pox after the small-pox, either after casual contagion, or inoculation? I have collected a great number of such cases, but want more.

R. DUNNING, ESQ.

MY DEAR SIR,

It is a long time since you have written to me. Why did you drop your correspondence? You positively must write, if it is but a scrap as short as this, just to answer my question; for really I have not the most distant guess to give at your long silence. If I calculate right, it is now near six months since I received a letter from you.

Believe me, yours very faithfully,

E. JENNER.

Cheltenham, February 21st, 1806.

P. S. I hope you and Mrs. Dunning and your family have escaped the long prevailing epidemic. It has fallen heavily on this house, and particularly on myself. Here the influenza assumes a character that I may call typho-catarrhal. It has confined me for near three weeks. What havoc the anti-vaccinists have made in town by the re-introduction of variolous inoculation! It is computed that, since April last, not less than 6,000 persons in the metropolis, and the villages immediately in contact, have fallen victims to the small-pox. One would scarcely conceive it possible; but these murders are, for the most part, to be attributed to the absurd productions of Moseley, Rowley, and that pert little Squirrel, to say nothing of Goldson. It is about London that the venom of these deadly serpents chiefly flows. So little have the people around me (though only 100 miles from it) felt it, that, since August last, I have vaccinated within a few of 1,500; and I certainly must deem it a piece of extreme good fortune that, out of the many thousands I have vaccinated, no failure or accident of any sort has arisen to my knowledge. Did you see a paper in the last journal from a Dr. Wood? I think it capable of doing great mischief, as it will tend to make practitioners careless about a point of great consequence, namely, an herpetic state of the skin coincident with vaccination, which you, as well as myself, have not only observed, but publicly and very properly noticed. My communications from various parts of the world are very cheering—800,000 cases from India. Adieu!

I beg to present my best regards to my medical friends of the Vaccine Society, among whom I have now the pleasure of claiming several acquaintances.

To R. Dunning, Esq.

My dear Sir,

What a propensity there is in all human beings to follow examples. You now behold in me a sad illustration of the fact; for I feel myself possessed of the same spirit of procrastination, with respect to answering letters, which has lately seized on you.

But I am the more reprehensible, as the inclosed paper ought not to have been so long detained. I can only say, by way of apology, now, that I meditated some other mode of conveyance, and that it goes from me with my hearty good wishes.

The above was written before your letter of the 17th arrived, which has at length found me at my cottage at old Berkeley.

A pretty sharp philippic, my good friend! but in such veneration do I hold the man of feeling, that if it had been ten times as sharp, I should have read it; though not without emotion, yet certainly without a murmur. Allow me just to make one observation. Should anything like the present occurrence ever happen again, let me entreat you not to indulge for a moment a fanciful speculation against your *friend*. As such I hope ever to be, and so to be considered by you.

I was happy to find you had been corresponding with Mr. J. Moore. He is an excellent man, and has produced an excellent book. I presume you know that J. Moore is brother to the general.

The impertinent interference of herpes with our vaccine pustule, I thought of so much consequence to be generally known, as to induce me to reprint my paper on that subject for distribution. With this you will receive two copies; one of which, I must beg your acceptance of; and the other you will have the kindness to present to the Dock Jennerian Society, with my grateful respects to the members. You will do much public good by enforcing attention to the progress of the vaccine pustule. If it be torn to pieces, either by the nails or the lancet, before the business for which it was placed upon the arm be accomplished, it is unreasonable to suppose that perfect security can follow. But to what purpose shall you or I address the public on these subjects, while such unprincipled characters as Moseley, and those who enlist under his banner, still continue to instil, or rather to push by violence, into the minds of the British nation their horrid doctrines? Have you seen Moseley's last pamphlet, the one just published? It is far more violent than any of the preceding. In this he has brought forward a string of cases, to point out *my* failures in vaccination—cases of small-pox after the cow-pox. But mark his audacity. They are of children I never saw in my life, and whose names I never heard of till they were placed before me in this murderous publication. Mr. H. Jenner, whose name he brings forward with a list of failures annexed, assures me that *the whole* is a most impudent forgery. What can be done with such a man as this? A general manifesto, with the signatures of men of eminence in the profession (and I really think we should now embrace nearly, if not quite, the whole), in favour of vaccination, would, if anything could, crush the hissing heads of such serpents at once; and I fear nothing short of it, unless parliament had a mind again to take the matter up.

I hope to be in London the first week in May, and shall attempt something; but, in the mean time, pray write to me, or I shall suppose my little packet has not reached you, or that your tremulous nerves do not yet vibrate in harmony with mine.

Long may you live to enjoy the sight, and your patients the benefit of your imperative R!

With every kind wish to Mrs. Dunning and your family, I remain, yours truly,   EDWARD JENNER.

Berkeley, 22nd April, 1806.

TO R. DUNNING, ESQ.

MY DEAR FRIEND,

For my credit sake, I hope you are not perfectly accurate in your calculation respecting the number of letters you have sent me without receiving answers. Forgive me if you are right; but really I believed myself sticking at the old number—one.

The perplexities I had to encounter, during my residence in London, last spring and summer, were of such a nature as cannot be described to you. My chief embarrassments arose from the vile machinations of that fellow, T. W.

Be assured I feel myself honoured by the request you and Mrs. Dunning make to me. Were I at perfect liberty, I would not ask for a proxy, but appear myself at your baptismal font, to take on myself the responsibility for young Edward's becoming a good Christian. Give him a kiss for me; and, at the same time, whisper in his ear that he has his godfather's best wishes for a prosperous journey through life.

By an intimation that has reached me from town, I find it is the wish of the College to be in possession of the reports of the faculty on the subject of vaccination as soon as possible. I mention this, that you and my friends about you may bestir themselves. I anticipate something energetic. Don't forget to attend to the last part of the inquiry of the College, namely, the cause which impedes the progress of vaccination in these realms.

For my own part, I do not scruple to attribute it chiefly to the industrious dissemination of the pamphlets of Moseley, Rowley, and the rest of the anti-vaccinists, which are calculated to excite horror and disgust at the very name of our admirable preservative.

As I am now addressing the secretary to the Dock Jennerian Society, perhaps he won't be offended at my asking for a copy of the Report about to go to the College. I have reason to suppose, from what has been already sent in, there is but one opinion there; and that that opinion is exactly what I could wish. I am sorry to put you to the expense of double postage, by sending you the inclosed Gazette, but I cannot procure a frank; and to withhold it from you would be almost criminal.

What a delightful narrative is here! what lover of vaccination can feel himself at war with his Catholic Majesty after its perusal! I must tell you that, from several countries where Balmis and his philanthropic companions touched, I have had most satisfactory accounts of the result. From Manilla and the Philippine Islands they send me an account of 230,000 successful cases. From Canton I have a most curious production; a pamphlet on vaccination in the Chinese language. Little did I think, my friend, when our correspondence first began, that Heaven had in store for me such abundant happiness. May I be grateful!

Present my best compliments to Mrs. Dunning and the family, and believe me, with best wishes,

Yours truly,
E. JENNER.

Cheltenham, December 10th, 1806.

To R. DUNNING, ESQ.

15, *Bedford Place, Russell Square,*
March 14, 1807.

MY DEAR SIR,

I have not yet heard whether your institution at Dock has made a report to the College of Physicians; I therefore begin to be impatient to hear from you, as the Committee will sit but a very little time longer.

Nothing transpires that is going forward within the College; however, the report to parliament *must* be luminous, and give fresh triumphs to my cause, and new laurels to those who like you have so ardently and so successfully engaged in it. I shall be necessitated to stay here for several months. When the days grow longer, and travelling pleasanter, perhaps you may take a flight to town. I will not despair of seeing you. You will be pleased to hear that the *dingy* Hindoo ladies are convincing me of their grateful remembrance, not merely by words, but by a *tangible* offering, while my *fair* Christian countrywomen pass me unheeded by.

If you should hear it reported that Sir George Dallas's

children, who were years ago vaccinated, have had the small-pox after variolation by Mr. Goss, put no faith in it. I have seen Mr. Goss's letter to Sir George, which sets the whole matter at rest. It has made no small buzz here. They are cases which lead to corroborate the millions that precede them of the efficacy of vaccination in securing the constitution from the contagious effluvia of small-pox. But no process can always secure it from the effects of variolous poison when inserted into the skin.

Believe me, yours truly,

E. JENNER.

RICHARD DUNNING, Esq.

MY DEAR FRIEND,

I have a thousand things to say to you on the vaccine subject, and all very pleasant ; but if I stay now to run into detail, the post will set off without my letter. However, it shall not go without a scrap or two respecting the Report of the College. This, I have every reason to suppose, contains everything that I or even my warm-hearted friends at Plymouth could wish for. Indeed, from what I have collected from some of my learned brethren, it far exceeds my strongest expectations ; for who could have conceived that, in so large a body, an opinion would have been *unanimously* given on the safety and efficacy of the vaccine practice ? But so I find it is. Among the immense mass of letters received by the College, several cases of small-pox after cow-pox, of course, were given. But, on casting up the numbers, lo ! it appears that the cases of failure were not so numerous as the deaths (according to computation) would have been, had the patients been vaccinated and placed in good hands. I should tell you that several of the Fellows began the investigation with no favourable impressions on the subject; but, as I have before mentioned, unanimity prevailed at the winding up of the business.

I have just received a note from the president, Sir Lucas

Pepys, requesting me to vaccinate his little grandson. Two years ago the worthy president would as soon have had the boy's skin touched with the fang of a viper as the vaccine lancet. But this *inter nos.*

I have some charming communications to make to you respecting tangible compliments I receive from Hindostan. This must be reserved until my next letter. Have the goodness kindly to remember me to my friends at Plymouth. They certainly honour me too much, but I shall ever hold them in grateful remembrance, and no man on earth more than yourself.

　　　　Believe me ever truly and sincerely yours,

　　　　　　　　　　　　　　　　E. JENNER.

London, 16th May, 1807, *Saturday evening.*

　　　　　　R. DUNNING, ESQ.

　　　　　　　　　　Berkeley, May 7, 1808.

There is not, my dear friend, within the circle of your acquaintance an individual who felt more poignantly for you and Mrs. Dunning, on account of your domestic affliction, than he whose name your little innocent bore.

Knowing as I do that the mind suffering under that kind of distress which you have so severely experienced can derive no comfort from condolence offered by any human being but by meditation, and that kind of intercourse with the Almighty, which is granted to all when sorrows overwhelm us, I had not the power to write to you. Do not, therefore, I pray you, attribute my silence to a want of feeling, but to its true cause, an excess of it. You had, and you still have, my sincere commiseration.

I was in town some few weeks ago, and had a conference with one of the ministers on the subject of vaccination, when I was assured that it was the intention of government to take it under its immediate consideration as soon as the more weighty concerns of the nation had gone through the House of Commons.

On this account I shall go again to town very shortly to assist in the arrangements, but of what kind they will be, I cannot at present inform you, but trust they will be of such a nature as to facilitate the progress of the practice. In my opinion a proclamation from the King, founded on the Report of the College of Physicians, universally dispensed, and recommended to the attention of the magistrates, the clergy, &c. would produce a striking effect. This would be greatly aided by allowing some pecuniary acknowledgment to those who vaccinated the poor. I have just received the annual report from Dr. McKenzie, superintendent-general of vaccination at Madras. Wonderful to relate, the numbers vaccinated at that presidency only in the course of the last year, amount to 243,175. From Bombay I learn the small-pox is there completely subdued, not a single case having occurred for the last two years. All my foreign reports correspond with these; but still Moseley, Birch, Pearson, and a few others, are using every mean and despicable artifice to keep up the prejudices of the people at home.

With every good wish, believe me, my dear friend, most sincerely yours,

E. JENNER.

To THOMAS PAYTHERUS, ESQ.

Berkeley, May 12, 1808.

DEAR PAYTHERUS,

I hope the people won't grumble at my conduct; but if they have anything to say against it, that they will speak out, then I shall know how to defend myself. I was called up in a mighty hurry in the winter, and when I came, no one seemed to know why or wherefore they had put me to this extreme inconveniency.

It was then agreed upon that there should be as full an assembly on the 17th as could be collected. I canvassed my friends, and indeed went so far as to prevail on some to accept the office of steward, but, lo! the three days appointed after my departure for the meeting of a board, when this matter was to be adjusted, were attended only by the secretary and his inkstands. Once or twice I believe there was the solitary exception of my friend John Ring. This was no wound to my feelings, considering myself abstractedly; but yet I think the measure injudicious, as it will give the enemies of vaccination a temporary triumph. I now hold myself at the beck of Mr. Rose. This he knows, by a letter I have lately written to him. From the business which, I perceive, by the paper of this evening, parliament have immediately before them, I don't think the vaccine business can yet be brought forward. With regard to my agency in the matter, I don't think it is worth a rush. Vaccination will go on just as well when I am dead as it does during my existence, probably better, for one obstacle will die with me—Envy.

You tell me I am accused of inactivity and indifference now, when I have received the parliamentary grant. Such language is unwarrantable, because it is unjust. I despise such scoffing; it is condemnation without a trial, and I shall hold all who thus reproach me in the most perfect contempt. If I could obtain a little peace and quietness, my pockets should readily restore every shilling they have gained by the cowpox discovery. That such a thing has been discovered, I, in common with the rest of mankind, have reason to rejoice; but this I also declare, that I wish it had been the lot of some other person to have been the discoverer; and in this wish I am sure my family have reason to join me very heartily; for they, as well as myself, are strangers, through it, to those domestic comforts which we should otherwise enjoy. So far from being inactive and indifferent, as accusers insinuate, that portion of my time which is not occupied by my ordinary professional pursuits, is entirely devoted to it.

To say nothing of other matters connected with vaccination, think for a moment what incessant labours my correspondence imposes upon me, labours which admit of no alleviation, as it must be done by my own hand only. Believe me, yours truly,

E. JENNER.

P. S. During my last visit to town, I called twice on Dr. Saunders; he was not at home; and once I wrote to the doctor on an interesting subject, but received not a line in reply.

TO JAMES MOORE, ESQ.

MY DEAR FRIEND,

I cannot possibly get the paper ready by to-morrow's post. Saturday is no post-day here; Sunday it shall be sent, if circumstances will in the meanwhile admit of my bestowing that time upon it which is absolutely necessary. I recommend extreme deliberation and circumspection in completing this important document. As for yourself, my dear friend, it cannot be expected that you can at present coolly exercise your correct judgment on any thing of the kind*.

The instructions sent out by the Royal Jennerian Society were framed by Addington, Ring, and myself, after some weeks' labour. Their basis was my original paper. It will be no disparagement to the Government Institution to avail itself of any thing useful that can be found in the wreck of the Jennerian Society; and I am inclined to think our paper a clearer and better thing than the manuscript you have sent me. I shall return it, with some alterations.

Though not as a matter of duty at present, yet be assured I shall be always ready, as an act of courtesy, to do any thing in my power to promote the ends for which the Government Institution was established.

Believe me, most truly yours,

EDWARD JENNER.

Berkeley, Feb. 9th, 1809.

---

* This was written soon after the fall of Mr. Moore's brother, the gallant and ever to be lamented general, in the battle of Corunna.

To JAMES MOORE, Esq.

DEAR MOORE,

Depend upon it there are many such cases as those which
have occurred in Mr. Wingfield's family in reserve for us.
Vaccination at its commencement fell into the hands of many
who knew little more about it than its mere outline. One
grand error, which was almost universal at that time, was
making one puncture only, and consequently one vesicle;
and from this (the only source of security to the constitution)
as much fluid was taken day after day as it would afford :
nevertheless, it was unreasonably expected that no mischief
could ensue. I have taken a world of pains to correct this
abuse; but still, to my knowledge, it is going on, and particularly among the faculty in town. Mr. Knight's cases
were first made known to me by Lady Charlotte Wrottesley.
This lady was one of my early pupils, and is an adept at
vaccination, as thousands of her poor neighbours in Staf-
fordshire can testify. She saw at once the true state of the
children in question. I do not presume to say, that these
children are examples of any improper practice; they might
have been affected with herpetic eruptions at the time of vaccination, which are so apt, without due attention, to occasion a deviation from the perfect character of the vaccine
vesicle. I think it must be the paper on this subject you
allude to as wishing to see. I have, therefore, sent it to you;
and a copy of that paper you saw in manuscript, on secondary variolous contagion. If you should want any more of
the latter, you may draw upon Gosnell the printer for them.
By the way, it might be right to send one to the National
Vaccine Establishment; determine this point yourself. Willan, in his Treatise on Vaccination, has spoken much to the
purpose respecting small-pox after cox-pox; you cannot
quote a better author. His word will go further than
mine, as he must be supposed to be less interested. I do
not think enough has yet been said of the small-pox after

supposed security from small-pox inoculation. Blair told
me, when I left town, he was collecting these cases with a
view to publication. Thousands might be collected; for
every parish in the kingdom can give its case. I fear your
materials for the year are more scanty than could be wished
for your Report; but they are in good hands to make the
most of. Addington will not be an improper addition to
your establishment. He has talents; and will be always
ready to assist you with his pen and ink when you are
hurried.

I am sorry to tell you our affliction still continues here.
Poor dear Edward still exists, but I shall soon be doomed
to hear his last sad adieu!

Farewell, my dear friend,
Yours most faithfully,
EDWARD JENNER.

In Willan's book you will find a letter of mine, part of
which may be interesting.

To JAMES MOORE, ESQ.

DEAR MOORE,

You are a good hand at a banter. The last five letters
I have sent you, I do not think you have read; I have a
right to suppose so, because you have not *answered* one of
them. Perhaps I may have received an equal number since,
but scarcely any thing I have said is noticed. All my great
and grand arguments are thrown away upon you; but you
tell of things that I have not said, " come to town if you
advise it." You cannot find this in any letter I have written, since necessity compelled me to put off my autumnal
journey. Let me tell you this said coming town is a very
serious matter. I have a large family to look after; and an
invalid wife, that is not a moveable commodity, and requires
my attention. I have a vast variety of weighty concerns to
look to at *home*. What home is, and what those concerns
are, you can have no conception of. You consider me as a

poor soldier on a furlough, and who must join his regiment at the beck of his officer. What does vaccination require of me now? If a new continent was vomited up in the midst of the great *Pacific,* and if it were peopled but with mermaids, I would then lend a hand for an arrangement to save them from the small-pox. If I am wanted, tell me what I am wanted for, and when.

I explained in conversation, as I said before, all that passed respecting my first paper on the cox-pox intended for the Royal Society. It was not with Sir Joseph, but with Home; he took the paper. It was shewn to the Council, and returned to me. This, I think, was in the year 1797, after the vaccination of one patient only; but even this was strong evidence, as it followed that of the numbers I had put to the test of the small-pox after casual vaccination. I should have sent you a copy of the petition to parliament before now, if I could have found it. If it was ever printed, it has escaped my recollection. I took care not to print it myself, as I was ashamed of it.

The newspapers announce a circular address on vaccination from the Secretary of State. If it goes to the magistrates, I shall have one of course; if not, pray send me one. Another is also spoken of as coming from the President. Did you see what the French minister says on the subject in his annual exposé?

To JAMES MOORE, ESQ.

MY DEAR FRIEND,   *Berkeley, Feb. 26th,* 1810.

I do not yet feel myself in that state of composure which will allow me to sit down, begin, and finish a letter to you; so I have thought it best to take a long sheet of paper, and fill it by little and little. By this stratagem you will hear from me in a reasonable space of time, and I shall not have to reproach myself for neglect.

I am much gratified at the thought of your having had an interview with my friend Hicks. He happened to be with me when your letter arrived; and I can assure you, that you are not more pleased with him than he is with you. Where there is a congeniality, there is often a friendship formed at first sight; and men unbosom themselves as freely as if their intimacy had been of long standing.

I have made a great blunder, it seems, in my reply to your inquiry respecting my opinion of what you call papulary* eruptions after cow-pox. I really thought you alluded to that appearance which I mentioned; but finding myself set right, I have no hesitation in saying, that what Willan has said on this subject is correct†. My friend Dr. Parry, of Bath, has made some interesting observations on these modifications or varieties of variola ; and I am sure he would readily furnish you with them on an application for that purpose. Creaser, of Bath, could also give you some good facts, with observations on the same subject. By the way, have you his pamphlet respecting P——'s bad conduct? You should have it. I have myself seen but one solitary case of this secondary small-pox, and that was in a child of Mr. Gosling's, vaccinated by a Mr. Armstrong, This went through its course in the usual rapid way.

You spoke of a print for your intended work. There are several about the town. The best, I think, is from a painting of Northcote's, done some years since for the Medical Society at Plymouth. I believe this is rather scarce; but you are acquainted with Northcote, and I dare say he has one in his possession. When I was last in town, my friends urged me to sit to Lawrence ‡, and I complied. If you approved of it, and he had no objection, that might suit you.

* Or was it secondary that you called them ? I cannot at this moment refer to your letter.

† The College, in their Report, have expressed themselves very well on this subject.

‡ From this portrait the print which adorns this work has been taken.

He talked of getting a print from the painting for himself. It will never do for me to go to the pencil now; for if my countenance represents my mind, it must be beyond any thing dismal.

I cannot refer to your pamphlet, as it is among my books at Cheltenham. If you have one to spare, pray send it to Harwood's, the bookseller, in Russel Street, who will soon send me some books from town.

Do you not intend mentioning cases of small-pox after supposed security from small-pox inoculation? Such cases are innumerable. I think there are thirteen on record among the families of the nobility. Blair, I believe, has collected the greatest number of them. You know my old opinion on the matter; that they occur, for the most part, through the interference of herpetic affections at the time of inoculation. One decisive proof you will find in Willan's vaccine book, given by me. From facts I go to hypothesis; and conceive that the appearance of the small-pox twice on the same individual arises from the same cause. On this subject I could write a long chapter; but as it would necessarily be theoretical, you would not thank me for it. I must just touch upon it. We see that variolous matter may be generated by inoculation on the arms of one person in that degree of perfection, as to communicate the small-pox by transferring it to those of another; yet the person, whose constitution shall in the first instance have been exposed to it, shall remain unprotected from future infection, although the system has been deranged during its presence on the skin. Where, then, is the difference, whether the morbid poison was confined, or limited to a point or two, or spread universally in the form of pustules? If the change required to give security could not take place in this one instance, why should it in another, under the same existing circumstances? The *peculiarity of the action* [I do not like to call it *morbid*, because it is generally salutary], is often too strong to be overcome, yet I am ready to conclude that this is not a frequent occurrence.

The more I reason upon it, the more I am convinced that the idea I broached in my first publication on the cow-pox, namely, that poisonous animal fluids are not absorbed and carried into the blood vessels, is correct.

I shall not send you any more Reports, till you tell me whether those already sent are of any use to you. Have you seen that lately published at Nottingham? It is the more valuable, because the Vaccine Institution there acknowledged a single failure, and this a positive one, among their vast number of cases. The Report from Ceylon I have before spoken of, but do not know whether you have seen it. You may get a copy at the Transport Office *. Perhaps, too, of your neighbour, Sir Walter, as it comes from his friend Christie. It should be made public. Mons. Corvisart, the favourite Physician of the French Emperor, has been good enough to present a petition from me in favour of two British captives, to which the monarch lent a gracious ear. In return, M. Corvisart requests me to use my endeavours to obtain the release of a young sea officer of the name of Rigodit (ensigne de vaisseau). He mentions him as residing at Wincanton in Somersetshire; but on inquiring of the Commissary there, I find there is no such person among the French prisoners at that place. Can you assist me in this business? It would be in vain to make an application to government without first finding him out; and when this is effected, I am somewhat at a loss to know where to make my application. I could have easy access to one of the royal dukes, if this would do.

John Gale Jones, I see, has at length succeeded in obtaining the situation for which he has long been a candidate. This fellow had once the impudence to desire a man to call

* I obtained it from thence under a promise of returning it.

on me in Bedford Place to say, that he, Jones, would advise me immediately to quit London, for there was no knowing what an enraged populace might do. He was the writer of Squirrel's book, the long anti-vaccine columns in the Independent Whig, and many of the most violent papers in the Medical Observer. I was held up in his Forum for several nights as an object of derision; but I silenced him by the same weapon as I have many others—contempt.

I send out a great deal of vaccine lymph on ivory points; but my stock is exhausted, and I am now reduced to bits of quills. Can you procure a supply for me; and a few papers of your instructions? Remember all parcels must come by the Gloucester mail coach. The combmakers who prepare the ivory points are apt to make them too small and narrow towards the extremities, unless instructed otherwise.

That Proteus Adams seems again to have become a vaccinist. The last number of the Medical and Physical Journal speaks aloud in its praises. The Report from Madeira is excellent, and I would wish you to refer to it.

I think you must now be tired of my long dull letter, so God bless you! I feel as if I never should be in spirits again. I cannot help but look back, and say to myself, if I had done so and so, I should not now have been enveloped in this dismal black cloud; but all this is very wrong; however, it is right that you should know every thing about me. Know, then, that I have been through life, almost from the earliest period of my recollection, haunted by melancholy; but yet, at times, my spirits have mounted to the highest pitch of vivacity. Whether they will ever rise again I know not. Adieu, my dear friend.

Our best wishes attend you all,

Berkeley, Feb. 28th, 1810.   EDWARD JENNER.

To JAMES MOORE, ESQ.

Berkeley, 21st April, 1810.

MY DEAR FRIEND,

I must get to my old plan of the long sheet and filling it up leisurely, or I know not when you may hear from me.

I must now thank you for your kindness in endeavouring to search out Monsieur Rigodit. There is something mysterious in this affair, as his friends in France still suppose him to be in England.

You must not rally me too hard about my theories unless you find them as wild and absurd as those of the persons you name. What I mentioned to you surely rested on something. They did not stand on a baseless foundation, but I placed them on analogy. If there were no theorists in the world, how slow would be the advance of science. Sydenham and Boerhaave were so impressed with their theories that they gave them to the world as facts. Not so your friend at Berkeley. I don't think you will ever catch me promulgating any theoretic notion which can positively be set aside by a fact. Such is the nature of the animal economy, that there are a thousand processes going forward which never can be stared full in the face; but there is no harm in a plausible guess.

What is your Establishment about? I fear little or nothing; but you will soon hear that a spirit of activity has shown itself in this county, which will do more to serve the cause of vaccination than any thing which has yet started up. Its advantages will be so self-evident, that it will soon run the kingdom over. You shall know the full particulars as soon as they come out. The great feature of the scheme is this, to place every man in a questionable point of view who presumes to inoculate for the small-pox, with such a mass of evidence as will be held up to him in favour of vaccination. A general association will be formed of all the medical men in the county favourable to the plan; and I really think, to avoid the ignominy of resistance, nearly the whole will come in. Some of the variolo-vaccinists have already abjured their old bad habits, and joined the standard before it was half hoisted.

I see my friend Hicks often; we don't forget you in our conversation. I don't wonder at your coming to a stand still in your *opus magnum*.* I hope it will be an *opus bonum;* but I think you have undertaken a tight job. The world is wide, and you have got to traverse over it. *Bon voyáge.*

You don't like my style when I write for the public eye, nor do I; but I cannot mend it, for I write then under the impression of fear; and it must be remembered, that when I write in London my brain seems full of the smoke. My great aim is to be perspicuous, and I got credit for succeeding in the papers first sent out; but some of the others might be more obscure through my taking greater pains with them: an error I shall be happy to avoid in future; for you know I am not fond of much work.

Truly yours, with best wishes to Mrs. Moore, and your family,

EDWARD JENNER.

P. S. How is your laughter-loving girl; and that fine boy with the philosophic head?

24th April.

I can bring clouds of evidence to support what I have advanced respecting the effect of cuticular diseases on the vaccine vesicle. This is certainly a subject of some moment; and before you go into any thing decisive upon it, I would have you enquire largely into it. There is a letter of mine upon it in Willan's book, written subsequently to the paper which I have circulated. Captain Gooch, in Brunswick-square, Foundling Hospital, bore testimony to occurrences at Cheltenham on a very large scale.

---

* This, and some of the following letters, allude to Mr. Moore's histories of Small Pox and of Vaccination, then in progress.

To JAMES MOORE, ESQ.

Berkeley, 20th May, 1810.

DEAR MOORE,

The Persian ambassador, I find, is about to take his departure from this country. It has often struck me, though I forgot to mention it in any of my former letters to you, that he should carry home with him some knowledge of the beneficial effects of vaccination, especially as Persia has hitherto turned its back upon it. So it appears from a paper published in one of the East India pamphlets.* Your seeing Sir Gore Ousley would settle the whole business.

I write to you on the back of our resolutions, that you may know what we are about in this county; and remember, this makes two letters for one—rather an unusual occurrence with

Yours truly,

E. JENNER.

I hope soon to muster up resolution, and go to town for a few days. What have you done with Sir L——?

To JAMES MOORE, ESQ.

Berkeley, Nov. 23, 1810.

MY DEAR FRIEND,

It strikes me that the most effectual way of lessening, if not subduing, the opposition to vaccination, would be to obtain a mass of evidence on the subject from a district of some extent in the county of Gloucester. Berkeley might with propriety be made the centre, as most of the faculty in the surrounding parishes received their instructions from me, and as the practice is of longer standing in these districts than in most parts of Britain.

* Communicated in a letter from Dr. Milne to Dr. Anderson at Madras, dated Bushire, March 13, 1805. If you have a friend at the India House, you may get the vaccine pamphlet there, published in India.

The result of such an inquiry would be most favourable. I will venture to say, from reports already made to me, it would appear that the decrease in the deaths by small-pox has been in proportion to the general or universal adoption of the vaccine practice. But how shall this be brought about? I could certainly accomplish it myself; but yet I think it would have a more fair and candid look, and form a stronger feature in the next Report of the National Vaccine Establishment, if it were done by the Board. Having obtained a list of the medical men, such questions might be addressed to them as would bring out the evidence required; such, for example, as the following:—

1st. When did you begin to practise vaccine inoculation?

2nd. What number of persons have you vaccinated?

3rd. What is your opinion, from the result of your own practice, of the preservative effects of vaccination against the infection of the small-pox?

4th. Have you found the vaccine disease to be injurious to the constitution?

5th. What is the longest interval between the vaccination of any of your patients and their exposure to the contagion of the small-pox without feeling its effect?

6th. Has your residence been long enough in ———— to enable you to form an estimate of the mortality occasioned by small-pox before you began to vaccinate, and since that period commenced?

In addition to your answers to these questions, the Board of the National Vaccine Establishment would be happy in being favoured with any general observations you may have to offer on this subject.

I wish you would, with my best compliments and wishes, lay these hints before the Board; and if they think the plan eligible, I should be happy in affording them every assistance in my power to render it effective. Here, then, is the logic of the thing:—If vaccination is found capable of making a deep impression on the accustomed ravages of the small-pox in one widely extended district, why should it not equally so in all?

I shall return again to Cheltenham on Tuesday next, where I shall be happy to hear from you, particularly on the subject of my former letter.

Believe me,
Yours truly,
EDW. JENNER.

To JAMES MOORE, ESQ.

MY DEAR FRIEND,

Another scrap.—The postscript of your last letter, written on the cover, escaped my notice till this morning, when I happened to take it up. The complaint against your points is partly just and partly unjust. Knowing that four times out of five, those I send out will pass into the hands of a bungler, I give them a double dipping; that is, when one coating is dry I dip again; and when they are to go beyond the seas, I put on a third coating. By this means I very seldom hear of failure from the points. However, I am persuaded this would rarely happen even from a single dip, if the operation was conducted properly. (You should open a vaccine school.) 1 once had some vaccine points returned to me that one of my countrymen had been poking an arm with, and found inefficient. After all this, with the self-same tools I did my work completely, finer vaccine pustules were never fashioned. I wish you would send me four points prepared in the usual way. I will use them and tell you the result. In general, the *doctors* don't make a puncture sufficient to admit a due length of the point; they bring its extremity only in contact with the wounded cutis.

Now to another point. I wish you could see Paytherus, and confer with him about Woodville. He will recollect some awkward things, for which, in rougher hands than mine, he would have been worried a little. But I must tell you what it was that kept my resentment in check. My first acquaintance with Woodville commenced at this place about twelve years ago. It happened that he was here during an excursion I made from hence to Berkeley, and in the interval, Woodville attended one of my children who had been seized with a violent fit of illness. On my return I found the child recovering, and felt so pleased with the manner in which Woodville had treated him, that although his conduct towards me in town called aloud for chastisement, yet I was restrained from obeying it, through the recollection of this event. When I found him about to publish his pamphlet relative to the eruptive cases at the Small-pox Hospital, I intreated him in the strongest terms, both by letter and in conversation, not to do a thing that would so much disturb the progress of vaccination, and finally prove so injurious to himself. Cases were shewn both to myself and nephew, the Rev. G. Jenner, (who was brought up to the medical profession) at the Small-pox Hospital, of patients covered from head to foot with pustules as correct as if they had actually arisen from contagion, or been produced by inoculation. Still no argument would bend him, and we found his assistant, Wachsell, equally inflexible. However, not many months after his book appeared, he came to me in Bond-street, where I then lived, and told me he had seen his error, and should publish his recantation, and dedicate his pamphlet to me. We parted, as I thought, friends; I thanked him for his liberality and kindness, in offering me the dedication; but how greatly was I disappointed when he sent it. Instead of finding generous and manly sentiments, it was in reality a satire. Do pray see Paytherus; he will give you a thousand odd anecdotes, and don't forget to ask him for his book on vaccination. He must not omit telling you what once happened at a dinner at Coleman's.

I am happy at hearing your letter has stirred up a proper spirit among the *Fins*. The bustle there must be great. What you allude to in my publication respecting *London air*, was an observation made in consequence of what Woodville and Pearson had represented to me in the country, not from my own ocular demonstration. Hearing them describe the progress of the vesicle on the arm, and the subsequent state of the skin as being so very different from my own vaccination in Gloucestershire, and knowing from actual experience, that erisypelatous affections in town assume an aspect very different from that they put on in the country, I thought it possible this might account for it; not dreaming of the blunder that had been committed. You must not forget to refer to a paper (I think in the Med. and Phys. Journal) of Pearson's, in which he sagaciously accounts for the appearance of pustules on chemical principles.

The inclosed paper you will find of no small value. Present it, if you please, to the Board, to be deposited in their vaccine archives. It was sent to me long since from Paris.

I have much more to say to you; indeed, your first letter I do not consider as yet answered; but now I must go to bed, or drop upon my paper. Excuse this sleepy letter.

Truly yours,

EDW. JENNER.

Cheltenham, twelve o'clock,
Wednesday night, 19*th December,* 1810.

I shall give you some trouble soon in assisting me to liberate a French officer, the brother of Husson (see the list of names in the inclosed paper) who has nearly lost the use of his arm.

To JAMES MOORE, ESQ.

MY DEAR FRIEND,

I send this scrap by a person going to town, to say that I will write to you *fully* in a few days.

As accuracy is the life and soul of history, I did not like

to answer an important question in your last letter save one, until I had applied to my nephew, the Rev. G. Jenner, who edited the pamphlet containing "the evidence at large." I am expecting to hear from him early next week, and his answer shall be sent to you. As far as my recollection now serves me, this evidence was obtained from one of the Clerks of the House of Commons, who always sat at the right hand of the Chairman, and took down the words of those who were called in by the Committee for examination. Finally, a fair copy was presented to me, and I certainly did not have it published without authority. Your queries shall all be answered in due order.

You must be tender on the subject of the R—— business.

Truly yours,

EDW. JENNER.

To JAMES MOORE, ESQ.

Berkeley, February 15th, 1812.

MY DEAR SIR,

You have received Sacco's book, I hope, which was sent by yesterday's coach, directed for you at the National Vaccine Establishment, Leicester Square. Pray present my respects to the President and the Board, and tell them that this or any other publication in my possession on the vaccine subject is much at their service. I am well convinced they are alive to the interests of the important cause they are engaged in; and it should not be forgotten that Mr. G. Rose assured me, (at the time, by the way, when I had no conception of not being a Member of the Institution,) that if more money was necessary for the completion of its objects it should be granted.

I rejoice at seeing so distinguished a person as Sir Francis Millman at the head of vaccine affairs. We wanted firmness and decision, and I now see that we shall have it. I beg you to present my best compliments to him, and to say, that when I go to town I shall have the honour of waiting upon him, and hope he will indulge me with a full conversation on the subject, particularly that part of it which relates to the conduct of the first Board, the cause of my seceding, &c. &c.

I could wish Sir Francis to see the manner in which our clever neighbours the French have organised their Vaccine Institutions. You will find it inserted in one of Bradley's Medical and Physical Journals.

My friend Ring would furnish you with it, if you do not know where to put your hand upon it. Ring translated the paper. It is very gratifying to me to see so great a decrease in the mortality occasioned by small-pox. But why should any one perish by this disease here, when there are so many examples before us of its being rooted out in every town, city, and district where vaccination is practised universally. Even in many of the populous districts around me, where large manufactories are carried on, and where the people with one accord have taken up the practice for the last ten or twelve years, the small-pox has been scarcely known during that period. I have often expressed a wish to the medical men of this country to report these facts to the Board. They all promise, but I believe few, if any, perform. A few lines, addressed by the Board, would have a strong effect, and the intelligence they would obtain would be very impressive in a future annual Report; and I submit it to their consideration whether it would not be more so than coming immediately from me. I assure you it was with some reluctance I sent copies of the Spanish papers. I must endeavour to do away the charge of egotism, by requesting you to consider that it is on vaccination they so lavishly pour forth their praises, and not on me. I have not received any late report, either from De Carro, or Professor Odier, at Geneva. Dr. Marcet, I believe, corresponds with the latter; and if any report has been inserted lately in the foreign journals, he will be very likely to furnish you with it. Professor Avelin, of Berlin, is, I believe, a character well known. From this respectable

man I lately saw a report in its way to America. I shall
subjoin a copy of its leading features. It was addressed to
Dr. Smith, at New York.

"The anniversary of the invention of the cow-pox inoculation, or the Jennerian Feast, was celebrated very solemnly
at Berlin, on the 14th of May last. By public accounts it
appears, that there were inoculated in all the Prussian
States,

In 1801	. . . .	9,772
1802	. . . .	17,052
1803	. . . .	50,054
1804	. . . .	102,350
1805	. . . .	43,585

"At these times the population was about 9,743,000.
From 1806 to 1810 (since the horrible war and the diminution
of the population to 4,338,000), the inoculated were 160,329.
Dr. Bremer only, at Berlin, in the Royal Institution for cow-
pox inoculation, had inoculated 14,605. The total, as officially and voluntarily sent to Government, amounted to
402,720 vaccinated, but certainly one-half was not officially
mentioned. It may certainly be at least 600,000, or even
800,000."

I shall have much to say to you in my next letter respecting my present state of retirement; in the meantime, believe
me, with best compliments to Mrs. Moore,

Very sincerely yours,

EDWARD JENNER.

P.S. Should not John Ring's station, which is so popular,
be incorporated with those of the establishment?

To JAMES MOORE, ESQ.

Berkeley, June 20th, 1812.

MY DEAR MOORE,

It is a long time since I heard from you, and longer still
I fear since you heard from me. What is become of the

Annual Vaccine Report? Have the hurley-burleys of the state annihilated it? If it exists, pray let me see it. If the South American and the Havannah Reports, with which I furnished the Board, are not noticed, those who sent them to me would think themselves not attended to with due respect. Poor Sacco and the seeds! This is a bad story, and I am in a scrape. Sir Francis, perhaps, did not apply to Sir Joseph; or, if he did, a request of mine was thought but little of.

I am still leading a sort of pastoral life here, and time flies on without leaving any thing behind it for *my biographer*. But I really do intend going to town, and then you shall see what you can squeeze out of me.

I have always thought that the subject of vaccination should be kept before the eyes of the public by means of the newspapers. This was never well done, and now it is scarcely done at all. Can you stimulate the Board to think of this? It would be very easy to give extracts from reports.

I have very lately received from Italy a Poem, " Il Trionfo della Vaccinia, by Gioachino Ponta," who, I hope, is a bard of celebrity, for he has spun it out to between 4000 and 5000 lines. It is beautifully printed, at the famous press of Bodoni at Parma.

Knowing nothing of the language in which it is written, it lies before me in a tantalising shape. I shall bring it to town. If it is a good thing, cannot we transform it into English?

Adieu, my dear friend, with best compliments.

To JAMES MOORE, ESQ.

Berkeley, August 9th, 1812.

DEAR MOORE,

This scrap, I fear, will be dear at the money it will cost you, though the sum will be but two-pence. I could not let

slip the opportunity of thanking you for your last letter, giving
me a month's respite from transportation from this place to
town. Certain it is, I have no society here but clods; but
out of those clods I contrive to make something. The produce of their fields has been a plentiful source of enjoyment
to me. This year there has been more of liver disease
among sheep, cows, oxen, hogs, and some other animals,
than I ever remember. I long since discovered that the
ordinary source of scirrhus is the hydatid, when passed on
to its secondary stage; but there was another sort of scirrhus
which puzzled me till now, and I make this out to originate
in diseased bile-ducts. Some of these I find dilated to the
size of a child's finger, and passing in this state almost to
the extreme edges of the liver; their internal coats highly
inflamed, like a croupy trachea, and throwing out mucus and
coagulable lymph. Others, which have weathered the in-
flammatory stage, thickly incrusted over with stony matter.*
Here, then, is a little apology for seclusion in this sequestered corner of our island. It is a singular thing, that the
liver itself (that part of it which remains unabsorbed),
should suffer the intrusion of any of these foreign bodies
(I may call the scirrhus hydatid foreign), and not be in the
least diseased, even the parts in immediate contact. Portions of the organ are taken away merely to make room for
these odd visitors. Enough of this for the present.

I shall say something on the *Report* in my next. That
part of it which points out the happy results of vaccination
among our troops must make the country feel, if they have
any feeling in them. I am hurt to think the small-pox again
rages. That must be the case, till inoculation is conducted
in a different way, if conducted at all. It does not appear
in vaccinating districts; for example, in this. As no particular notice has been taken of the foreign communications, I
am thinking of sending them to one of the periodical jour-

* I do not call this scirrhosity, but it produces scirrhus.

nals. The Edinburgh Quarterly Journal is the most respectable.

You do not seem to have understood me clearly respecting *newspapers*. It would certainly be *infra dig.* to go into controversy; but not so to lay cheering and persuasive reports before the public through this widely flowing channel. This is what I meant, and I hope you will agree with me in the propriety of the measure.

Make my affectionate regards to Mrs. Moore, and believe me,

Truly yours,

EDWARD JENNER.

TO JAMES MOORE, ESQ.

Berkeley, October 11, 1812.

MY DEAR FRIEND,

When you wrote last, about a month since, your accompanying the life guards to Spain was a point not absolutely fixed. How stands the matter now, my friend? I hope they will not be so cruel as to send you there. The event I have long been expecting has at length taken place. I have lost my only sister, and the last of my family of that class : so that I am now insulated in that way—the only one left of ten. I could not come to town while she lay on her death-bed; but now I shall come, and try to cheer myself, by mixing with my friends for a week or two; that is, if you think it would be a good time, which I doubt, for some tell me the town has no company in it. Write soon, and your letter shall be my warrant.

The inclosed I received a short time since from Professor Waterhouse at Boston. There is something so striking in it with regard to the politics of America, and so unlike what we are taught to believe at home, that I have inclosed it for you, thinking it might be of some value in the hands of some of your political friends. I know you are well acquainted with my Lord Lauderdale, and many others. It may be sent to a newspaper if you think it may be useful; in that case, no names must be mentioned. Dr. Waterhouse is a man of correct habits. For the seven first years of vaccination I corresponded with him regularly. He upbraids me justly for late irregularities.

I have not heard lately whether the fury of the small-pox is abated in town. I trust it is. Had I power to exercise vaccination as I liked, in one fortnight this dismal work of death should entirely cease. What a sad wicked fellow is that Birch. Moseley I hear nothing of now, but Birch is still employing his agents to spread the pestilence.

To James Moore, Esq.

Chantry Cottage, Berkeley,
November 18, 1812.

My dear Friend,

Yours of the 21st of October, I perceive still remains unanswered. The truth, and nothing but the truth, of the matter is, that I intended to have been in town before this time; but, as in former days, was prevented. It is of no use, I know, for me to make excuses to you, for you will uncharitably think me a wilful procrastinator.

There is a vast deal of wisdom in the two little words selected by the sages of antiquity as a guide to medical men and others. " Festina lente." What a jewel of an axiom ! But then, my friend, observe, I make no encroachments upon its meaning when I take it as a rule of practice.

Since I wrote last, I have received from St. Petersburgh a report of the progress of vaccination throughout the whole Russian empire. It is very copious, and very interesting. Such documents should go before the public; therefore, I do not see the propriety of sending it to the tomb in Leicester Square, where lie interred, without a record, numerous branches of the same family, most of whom were born and

LIFE OF DR. JENNER. 383

bred beyond the seas; some in South America, others in the
West India Islands, and elsewhere.

I should much like to see your paper containing the History of Vaccination, and the exploits of the man who brought
it up. In looking over my papers, I have found a great
many which will throw a strong light on the conduct of Dr.
P. Is there any chasm in this part of your history? It is a
very important part, and justice demands the exercise of
severity. It must begin with the Petworth business. This
is given by Lord Egremont. Next his uniting with Woodville, and forming (without mentioning the matter to me) his
institution. His cajoling the Duke of York to be patron.
The Duke's disgracing him. His spreading the small-pox
through the land and calling it the cow-pox, explaining *chemically* the reason why it had changed its character. His
treatment of me before the Committee of the House of Commons, attempting to prove that there were papers found in an
old chest at Windsor, which anticipated my discovery. The
portrait of the farmer from the Isle of Purbeck, with the
farmer's claim to reward, as the discoverer at the foot of it,
with a thousand minor tricks; and finally, finding all trick-
ing useless, his insinuations that vaccination is good for
nothing. The *Anti-Vacks* are assailing me, I see, with all the
force they can muster in the newspapers. The Morning
Chronicle now admits long letters. Birch has certainly
much the worst of it there. Can you tell me who my friend
and defender is in the Sun, who signs himself Conscience?

Do you ever see anything of your neighbour John Ring?
He writes but seldom to me now, and when he does write, it
is not in his old pleasant strain. Nothing is going wrong with
him, I hope. I wish you would find out; for, with all his
peculiarities, he is an honest fellow, and I have a great
regard for him. He has been paying money for me to some
of the institutions, and the inclosed draught, if you would
have the goodness to take it to him, would be an excuse for

your calling on him. I have made it payable after ten days, as this will approach you in a round about way.

I hope you will tell me in your next letter, which you must write soon, that the bills of mortality no longer hold up such a long list of slaughtered victims as I saw some time ago.

Rigby, of Norwich, a medical man well known in the world, contrived to stop the havoc in that city most expeditiously by an ingenious contrivance. He is the acquaintance of Charles Murray, and sends him his papers. The one he sent to me on the subject I now allude to, was written in the early part of the present month, and comes out as an Appendix to his other papers on medical police. Sir Francis should see it. My letter is getting too long; I must stop, and only add my best wishes to you and yours,

EDWARD JENNER.

I hope Sir H. Davy will not lose his eye from his late accident.

To JAMES MOORE, ESQ.

Chantry Cottage, December 17, 1812.

MY DEAR FRIEND,

I must animadvert presently on some parts of your last letter; but first let me thank you for your very kind attention to my nephew E. Davies, from whom I have heard by this evening's post. You have quite fascinated him. He speaks of you in such terms that I must not repeat, *as you are so given to blushing.* I may venture on one extract.

" Mr. Moore read to me a very considerable part of his intended publication. The style, in my opinion, is admirable; nervous, concise, gentlemanly, and severe without descending to scurrility. It is also so amusing as to render it interesting to every class of readers."

I desired him to tell you that a few days since I received a Report from the Deputy Inspector of Hospitals at the Cape of Good Hope, Dr. Hussey, of the annihilation of the small-pox, which appeared there in one of its most horrible forms, by means of vaccination. There may be no necessity for my sending it, as I find the National Vaccine Establishment is in possession of a similar document, or at least the purport of the communication made to me. However, as mine may go more into detail, 1 beg you will present it to the Board, together with the Russian Report. But I must again entreat you to request Dr. Hervey to see that they may be restored to me on demand, for I hold these things as sacred deposits, and they will pass from me as heir-looms. How different are these, my dear friend, from those trinkety baubles which mankind in general are so proud of transmitting to posterity.

Lest the Rev. Mr. Reed should not have laid before the Board a copy of his last edition of a very impressive paper, I shall put up one. 1 wish Sir F. Millman would recollect that upwards of seven hundred reports in favour of vaccination lie buried among the archives of the College of Physicians. Have you the Report from the Mauritius? I hear it is of the same nature as that from the Cape. It has not yet reached me.

The intended proclamation will put us all in battle array. We shall have sharp work for a little time, and we must be prepared with troops in the House of Commons; but our great guns will make such reports, that our enemies there will be stunned and astounded; even *Sir Francis* himself, who has been heard to say, that " cursed was the day on which vaccination was discovered."

In the midst of these reporting times, pray do not let Dr. Christie's be forgotten. Among all the good ones, there is nothing surpasses this. Lest he should not have sent it to the Board, I will write to him for that purpose. If you want *Home* Reports, send to Manchester, Birmingham, Chester, and other populous manufacturing towns, but above all to Norwich, from whence I have very lately received one *exactly* of a description with that of the Cape. It is printed, and was drawn up by Rigby of Norwich. Charles Murray could easily get this, and, believe me, it is worth your having.

Having written so much, I must defer my Philippics for the present.

Let me hear often while these important movements of the Board are going forward.

Adieu, my dear friend, very truly yours,

E. JENNER.

One line I must beg on the arrival of the packet, that I may know of its safety.

TO JAMES MOORE, ESQ.

DEAR MOORE,

After so long a pause in our correspondence, your letter was a high treat to me. How this happened, I will not take upon myself to determine. Doubtless, you thought me in fault, and I thought you; but perhaps the truth is, we were both so, and under that impression I quit the subject. I was ill almost the whole spring, and now, though better, am far from well. My nerves are in such an odd state, so exquisitely tuned, that unless they are touched by the most delicate finger, one who knows the *instrument* perfectly, in an instant all is discord. You only want a little more practice, and then no one would play better on it than yourself. Did you see my letter to Charles Murray? I hope you did, as I there expressed my opinion of your last report in high terms of praise. If this does not silence the malevolent tongue of opposition, what can? Yet, how wonderful, while this convincing document was lying on his table, that a great Law Lord should have so exposed himself among his brethren. "Vaccination did not merit the high encomium passed upon it; it was very well for those who, like himself, brought up their families in a large city, and was a security for eight or nine years." This language, I assure you, has made a serious impression in many parts of the country; for the people, who do not reason, conceive this exalted character knows every thing. There is another Bill, I see, brought into the House by my Lord Boringdon. Can you contrive to send me a copy when it is printed? I hope it does not come from the same bad source as the former.

Believe me, I am willing, and shall be always ready when able, to assist you in your literary toils. It would be very ungrateful in me if I were not, as you may say, De te *historia narratur.* If you will once more place before my eyes any materials you stand in need of, they shall be forthcoming should it be in my power to produce them. An old associate of mine has long been threatening to send some memoir into the world, but I have been constantly intreating him to desist, conceiving, that independently of the vaccine discovery, there was nothing of sufficient interest to engage the attention of the public. Believing that I have succeeded, I think he would have no reluctance in furnishing you with his scraps if you thought it worth while to apply to him. You need not mention this intelligence as coming from me. His name is Edward Gardner, and his residence is Frampton, near Stroud, Gloucestershire.

You speak of searching the British Museum for facts and opinions about the small-pox. Would you not find more by searching the publications of Woodville and Haygarth? I believe they have both given its life and adventures, and I trust it is left for you to record its death.

I like your little essay on poverty and riches. Give me (as a good man in the Scriptures said) neither one nor the other, but wherewithal to be content. I know you fancy that the cow has fattened me, and that it is of no use for me to attempt altering your opinion. My state of domestication is the same now as it was before I cultivated her acquaintance so closely, except, that then I had horses to my carriage, and that now I have none, and precisely for the same reason as should govern the conduct of all prudent men. To know any thing about me you should come down and inquire of my neighbours what I am, and what I was. Then, perhaps, your quotation (quantum mutatus ab illo, &c.) may still well apply, but not exactly in the way you intended it.

In one of your letters you seemed not perfectly satisfied that the fact respecting the origin of the vaccine was clearly made out. For my part, I should think, that Loy's experiments, independently of my own observations, were sufficient to establish it, to say nothing of Sacco's and others on the continent. However, I have now fresh evidence, partly foreign and partly domestic. The latter comes from a Mr. Melon, a surgeon of repute at Lichfield. He has sent me some of his equine virus, which I have been using from arm to arm for these two months past, without observing the smallest deviation in the progress and appearance of the pustules from those produced by the vaccine. I have at length found the French document I formerly alluded to, which with Melon's, shall be sent to you in the course of the ensuing week.

Allow me to congratulate you on the promotion of your meritorious brother, and to assure you that I take an interest in every thing in which your happiness is concerned; so be assured of my regard, and believe me,

Most truly yours,

E. JENNER.

Chantry Cottage, July 23, 1813.

To James Moore, Esq.

DEAR MOORE,

My friend and neighbour, Mr. Hicks, will deliver to you the promised papers respecting equine virus. I have been constantly equinating for some months, and perceive not the smallest difference between the pustules thus produced and the vaccine. Both are alike, because they come from the same source. If he does not give you a good scolding for your horrible letter in the spring, he will not be faithful to his commission.

I hope Murray has shown you my letter respecting Dr. Baron's intended publication.

Truly yours,

EDW. JENNER.

Berkeley, 1st August, 1813.

To JAMES MOORE, ESQ.

MY DEAR FRIEND,

I have had so much intercourse with you lately by means of London visitors, that my being a letter in your debt almost escaped my recollection. You have doubtless seen Charles Murray since his return from Cheltenham. I had two days of his company, and we pretty well talked over London matters. It was not then known that your late excellent president was tottering on his vaccine throne, from which I find he has since fallen. This is very tantalizing, as he was in possession of that stock of knowledge which rendered him fit for his government. I am a little acquainted with your new chieftain, but want to know your sentiments of him. I have always considered him as a very worthy man, of manners extremely gentle. In the hour of necessity, however, I hope he will be firm; and if the first Lord in Parliament should offer to degrade vaccination by uttering an untruth, (as one of these dignified personages lately did,) I trust he will not suffer a remonstrance of so tame and insipid a nature to come forth as appeared in a late circular. This is the only flaw observed in the administration of Sir Francis. You must approve these animadversions, as they come from an " ingenious gentleman."

I have heard no more of John Walker or Joseph Leaper, since I sent my positive refusal to become an associate in their plans, which, from such men, I think could have no good in them.

You made me happy in saying you had seen those excellent young women, the Paytherus's, and learnt from them what an active life I lead when at Berkeley. How different and wrongly formed were your conceptions of me. I do not yet despair of seeing you there when I again retire. How you would enjoy seeing me in the exercise of my magisterial powers, dealing out my lessons of morality to the poor unfortunate daughters of vaccina, when exhibiting their untimely *prominences*. I bring them all to the altar with their swains if I can; but, perhaps, I do not better their condition much by this; for matrimony among the poor orders of the peasantry is in general a wretched state.

I long to see the progress you have made in your book. Is it impossible to bring it here? You may be in Piccadilly at seven in the evening, and your arrival at Cheltenham be announced by the horn of the mail-coach at ten the next morning. I am sorry you have not succeeded in infecting a cow. I have told you before that the matter which flows from the fissures in the heel will do nothing. It is contained in vesicles on the edges and the surrounding skin. Did I ever inform you of the curious result of vaccinating carters? These people from their youth up have the care of the horses used for ploughing our corn lands. Great numbers of them in the course of my practice here have come to me from the hills to be vaccinated; but the average number which resisted has been one half. On inquiry, many of them have recollected having sores on their hands and fingers from dressing horses affected with sore heels, and being so ill as to be disabled from following their work; and on several of their hands, I have found the cicatrix as perfect and as characteristically marked as if it had arisen from my own vaccination. Birch and Brown, of Musselburgh, I hear, still pursue me in the newspapers, but I do not seek after their essays; for really I think them now greater objects of commiseration than resentment. Moseley, I believe, is silent. Your last report should be perpetually going forth from Leicester-square. It will never be old, and a few spare pounds would procure a reprint. How goes on small-pox among you? I am almost afraid to ask; and afraid, too, to inquire about Lord Boringdon's intended Bill. There has certainly been ample time for its preparation. I think it a little strange that he should never have made any communication to me on the subject; the more so, as I am acquainted with his lordship, having vaccinated his eldest child.

I have some reason to think that all *etiquetical* impediments to my becoming a member of your Board will soon be removed. I dare not say more on this point now; but the mystery shall be unravelled in my next letter. Sad complaints about your ivory points; and so there must be till they are better fashioned. The chance of infection will be in proportion to the coated surface introduced into the puncture. Your points are now become almost as fine as needles. This is downright *tailoring;* and I hope for my own sake, who am so pestered with letters, that some reformation will take place in this department of the establishment.

You begin to yawn over my long letter, and so do I, for it is almost twelve o'clock; so adieu, my dear friend, and believe me ever truly yours,

EDWARD JENNER.

Cheltenham, October 27, 1813.

To JAMES MOORE, ESQ.

MY DEAR FRIEND,

I am always happy to hear from you. You paint the passing hours in glowing tints, and I, who believe in prophecies, am a firm believer in you when you predict that the amelioration of the world is at hand. Till now, our views of what the twenty years' commotion in Europe was to bring forth, were dim and obscure; but the "still small voice" has ordered the mists and clouds to be dispersed, and through a clear and serene atmosphere we see a beautiful order of things gradually rising, as it were, out of chaos. Let us be grateful.

You see I was not quite in so great a hurry as my friend Christie to shew myself at Carlton House. I shall be there in good time, you may depend upon it, and then hear your history of the rise, progress, and downfal of a monster still more horrible than Bonaparte. You delight me with what you say of the new Board; and I must now mention a circumstance which will put their activity and zeal a little to the test. You probably may not have seen a pamphlet lately published by Dr. Watt of Glasgow, as there is nothing in its title that developes its purport or *evil tendency*. "An Inquiry into the relative Mortality of the principal Diseases of Children," &c. The measles, it seems, have been extremely fatal in the city of Glasgow for the last four or five years among children, and during this period vaccination was practised almost universally. Previously to this, the measles was considered as a mild disease. Hence Dr. Watt infers that the small-pox is a kind of *preparative* for the measles, rendering the disease more mild. In short, he says, or seems to say, that we have gained nothing by the introduction of the cow-pox; for that the measles and small-pox have now changed places with regard to their fatal tendency. Is not this very shocking? Here is a new and unexpected twig shot forth for the sinking anti-vaccinist to cling to. But mark me—should this absurdity of Mr. Watt take possession of the minds of the people, I am already prepared with the means of destroying its effects, having instituted an inquiry through this populous town and the circumjacent villages, where, on the smallest computation, 20,000 children must have been vaccinated in the course of the last twelve years by myself and others. Now it appears that, during this period, there has been no such occurrence as a fatal epidemic measles. You would oblige me in making this communication to the Board, with my respectful compliments.

The preceding pages were written some days ago. I have since had a call into a distant part of the country, and luckily into the land of vaccination. The medical man I met has been near five-and-twenty years a practitioner in one of the clothing districts, consequently a part of the county where population swarms. He is ready to testify that Mr. Watt's doctrine will not find the least support in any part of the wide range he takes, and where vast numbers of children have gone through the measles just in the same way as if they had previously had the small-pox. I shall balance this unpleasant piece of information with something of an opposite kind. The University of Oxford, on Friday last, conferred on me the degree of Doctor in Medicine, by diploma, without a single *non placet*. This is the more honorable, as I understand they consider this gift so precious that it is not bestowed twice in a century. Some early day next week (Tuesday, most likely) I intend going to Oxford to accept this boon, and staying one clear day.

Now, my friend, what say you? Do you feel bold enough to face me there? It would be a high gratification, most certainly; and I would envelope you in a frank, for you have no business to jaunt about and spend your money.

You see what paternal care I take of you. By the way, would not some of the sages there aid your research in conducting you over the Bodleian library? There are several Oxford coaches go from town every morning. I have a thousand things to say to you. Pray inquire of Dr. Hervey whether I may not knock boldly at the door of the College of Physicians and gain admittance; and desire him to explain the nature of the ceremony that would take place.

If you can come to Oxford, write soon, that I may fix the day for certain. Bring, if you can, the last bill of mortality. I dread the sight of it.

Most truly yours,
EDWARD JENNER.

Cheltenham, December 6th, 1813.

P. S. John Ring has been in dudgeon, and broken off his correspondence with me near a twelvemonth. I have no conception why; I wish you could find it out.

To James Moore, Esq.

My dear Friend,

You must excuse everything I do amiss now. Your two letters have remained long unanswered, and I wish these were all, as it would relieve me from some of my anxieties; but what is to take from me my heavy load of sorrow? *That* you cannot tell. You have care enough of your own, and I will not entangle you into a participation of mine.

Can I afford you any assistance in your laborious work, the History of Vaccination? I am at present unacquainted with your plan. I suppose you will trace it step by step over the globe, and shew the little opposition it met with from professors abroad, compared with what it found from those at home. Certainly here, the opposition was marked with unexampled atrocity. Among the many, you will find a difficulty in fixing on the man who has a claim to the severest stigma. The sale of a part of the Small-pox Hospital delights me. It will be a charming feature for you.

I think your conjecture a fair one respecting the spreading of the small-pox in London. We know nothing of it in this district, nor have we for near sixteen years. The same may be said of Cheltenham and its vicinity, except now and then a straggler passing through the town; but then it was always insulated, or nearly so; and this, by the way, is the only stimulus* the common people feel for bringing their children to be vaccinated. This apathy should be roused. I like the idea of the goblet, and wish more of these things were distributed. What if £500 per annum were allowed by government, would the national purse find any diminution in its weight? Your countrymen fairly won the prize. A

---

* The accidental appearances of the small-pox.

similar present, I think, should be sent to some one in this country, or it will look like partiality. The reverend Mr. Reed, I think, has a claim : you have got his circular.

There is a lady whom I could name that has vaccinated 10,000. But above all, in this country, I think John Ring, with all his peculiarities on his head, stands foremost. Think, my friend, on his vast losses in devoting so much time and expenditure to our cause, and pray mention it to the Board. I should not regard paying for it myself, if it could be done without his knowing it.

I admire the ingenuity of your metaphor, but I must cut it down to ordinary prose. You talk of rekindling the lymph when its fire has gone out. Its quality may be so modified by passing through herpetic skins, that it becomes unfit for the intended purpose. It will produce pustules of a diminutive size, with a faint, or even without any, areola, and finishes its course prematurely. Is this what you mean to tell me ? I hope it is, for it is very important. When in the deteriorated state, it gets into bad hands, and much mischief may arise.

Now for the sable emperor. You speak of something inclosed in your packet from a negro gentleman who is going to Hayti. Nothing came, at least from him ; but there was a letter from Mr. Wilberforce, speaking of the wishes of this gentleman with respect to vaccination at Hayti. Pray contrive to present my compliments, and to assure him, the black gentleman, how much pleasure it would give me to do anything in my power to further his wishes. With this I shall send two or three detached papers of mine, which may be useful, and some others. They should be accompanied by my original paper, but it is out of print. Can't you get him to write to me ? I should like a letter from him very much. Indeed there is another reason for my wishing it. I have had the misfortune to lose or mislay Mr. Wilberforce's letter, so that I am ignorant even of the address of this enlightened African. I must now give you a little history in which you will hear something of Petion, the semi-sable Emperor of Hayti, who I understand divides the kingdom with Christophe. In my list of patients, last autumn, at Cheltenham, were several gentlemen of respectability settled as merchants at St. Domingo. One of them, a Mr. Windsor, informed me how much it was the wish of Petion to establish a regular vaccine institution there. I promised to furnish him with vaccine materials, but was prevented from what befel me at that period. Mr. Windsor took instructions for calling on the National Establishment, but as you say nothing of the matter, I don't imagine you saw anything of him. All the gentlemen whom I have seen from the island speak of Petion in the most exalted terms, as one possessed of great intellectual powers, and who employs them for the best of purposes. Now what shall we do in this matter? I must leave it to your discretion. Mr. Windsor's address was at Messrs. Peel, Turner, and Scott, 109, Cheapside : but I fear he is gone.

You ask me to come to town. The quiet of this place suits my mind much better at present. But I call into action all the reason I can muster, and have always company in my house. These privations are very dreadful, and make a man wish he never had existed; but wishes of this sort should be banished, and give way to patience and resignation. My daughter is with me, and begs her best remembrances to you and Mrs. Moore. Robert is at Oxford, and would be glad to see you if chance should take you there, at Exeter College. When you see Miss Dunbar, give her our best wishes.

Pray don't serve me as I serve you, but give me another letter soon.

Yours, my dear Friend, most truly,

EDWARD JENNER.

Berkeley, December 3rd, 1815.

Perhaps I have mistaken the whole business in sending any papers. Set me right if I am wrong. The paper on "the Varieties and Modifications" should have universal circulation.

To James Moore, Esq.

Berkeley, March 5th, 1816.

My dear Friend,

Our correspondence has again grown slack; no blame lies at your door, but all at mine. I should have told you before this time, that I feel cheered by what you said of the vaccine medal, and the poem which was found enveloped in so much splendour in the library of the ex-Emperor.

Mrs. Moore saw my copy of the poem, and I do not think liked it much. Perhaps she might think the thread spun a little too fine. The poet's fancy has certainly flown in all manner of directions, and if you would like to judge for yourself, my daughter bids me tell you she will with pleasure copy for you a faithful analysis presented to me by a lady here, a complete mistress of the Italian language. I do not mean the whole poem, but its outline. The fact, as you have an excellent knack at managing these things, would perhaps find admittance with some advantage in the work you are now engaged in, as a *rub* to the British Bards, not one of whom, whose voice has obtained celebrity, has sung one single note in honour of Vaccina. Anstey, perhaps, may be considered as an exception, who piped up a Latin Ode about a dozen years ago, which the indefatigable John Ring translated neatly into English verse.

You are no stranger, I dare say, to a murmur that is spreading through various parts of the Empire, excited by what has been supposed a deteriorated state of the vaccine matter. Much has been written upon it in the public journals, and much has been said to me in private correspondence. Medical men are more expert than any others in discovering causes without the fatigue of much thinking, and in the present instance they have all hit upon the wrong

one—no great wonder. They attribute the lessened activity
of the matter which may happen to fall into their hands, and
its disposition to produce imperfect vesicles, to the great
length of time which has elapsed since it was taken from the
cow, and consequently to the immense number of human
subjects through whom it has passed. This is a conjecture,
and I can destroy it by facts. The matter may undergo a
change that may render it unfit for further use, by passing even
from one individual to another, and this was as likely to
happen in the first year of vaccination as in the twentieth;
for in spite of long experience, and instructions sent out
from societies and individuals throughout the country, there
are still medical men who will take any thing they can catch
under the mere name of vaccine matter, or from a pustule
incorrect in all its genuine characters. To guard against
this important error, I have again and again pointed it out in
every way I could think of, and at the same time made remarks upon its ordinary source. It is, then, from the
spread of matter of this description through many districts
that the dissatisfaction I speak of has arisen, and I fear
there will be some difficulty in setting aside the delusion;
for alas! how much more easy it is to see what is right
and good, than to effect it. The matter sent out by
the National Vaccine Establishment is much complained
of. I was applied to a few weeks since, by the surgeons
of the hospital at Gloucester, for some vaccine matter,
and their request was accompanied by the following
observation: "that after using thirty points sent from
town, not a single pustule was produced." The fault could
not be in the mode of using them, for those sent by me were
effective. I vaccinate the poor here weekly, and the pustules
(vesicles, if you please) are in every respect as perfect and
correct in size, shape, colour, state of the lymph, the period
of the appearance and disappearance of the areola, its tint,
and finally the compact texture of the scab, as they were in
the first year of vaccination; and to the best of my know-

LIFE OF DR. JENNER.

ledge, the matter from which they are derived was that taken from a cow about sixteen years ago. If there were a real necessity for a renovation, I know not what we should do, for the precautions of the farmers with respect to their horses, have driven the cow-pox from their herds. If you find any thing here worth communicating to the Board, I beg you will present it with my best compliments.

What shall you have to report this session to Parliament? Your small-pox list is much longer than one could have wished, but it is pleasant to hear that the next year's account of the mortality promises at present to be far more satisfactory. I have not observed from any quarter, where comments have been made, and where this list has been called tremendous, that it has brought forth a comparison between the fatality of the small-pox now, and previously to the introduction of the vaccine. Justice to the cause demands this, and I hope it will not escape your recollection when you form your Report. We may surely calculate on the reduction of at least half the number; for, if I recollect right, the average amounted to two thousand annually. In the provinces, the reduction is *very far* beyond that of the metropolis.

All this was written before your second letter arrived, conveying the sad intelligence of your being an invalid.

I am no stranger myself to the sciatica, having had many sharp attacks. The last time was in London, four or five years ago, while I was living in Cockspur-street. One day in an agony of pain, I resolved on trying the popular remedy, walking; and effected a most painful piece of pedestrianism up to Temple-Bar and back. On my return I flung myself on my sofa in a state of exhaustion with torture and fatigue; but it really proved a cure, and I have never had a relapse.

Your friend, Lord Sidmouth, was once a friend of mine, and perhaps remains so still. His good humour was an over-match for his firmness when Premier. Such characters

are estimable, but not fit to take a lead in state affairs, no
more than my acquaintance Dr. Lamb is in the affairs of
the Board of Directors. I never think of this part of the
system of your establishment without irritation. As soon
as a set of men have learned how to conduct the business,
they vanish, and others are put in, who are totally ignorant of vaccination.

The account you give of your boys, is very pleasing to
me; for be assured I take a sincere interest in every thing
that adds to your comfort.

My coming to town this spring or summer is very uncertain. I cannot make a good report to you of my health.
Among other maladies brought on by my sad domestic
affliction, was palpitation of the heart. This at intervals
still pursues me, and a very unpleasant sensation it is, especially as it prevents sleep; but I am tolerably easy about
it, as at my time of life I must expect to see and feel the preparation going forward for the extinction of vitality: but so
long as it remains unaccomplished, I shall remain, my
dear Friend,

Most truly yours,
EDW. JENNER.

Berkeley, 21st March, 1816.

My best wishes attend Mrs. Moore.

To JAMES MOORE, ESQ.

MY DEAR MOORE,

Before you make a comparative calculation of failures
between the vaccine and variolous inoculations, you must
consider the immense disparity between the numbers inoculated with the one and the other. If you calculate on a
period of forty years, I should conceive that in the course
of the last twenty years there have been at least five times
as many vaccinated as have been variolated.

Then you must take into the account *failures* attributable to ignorance, neglect, &c. &c. &c. Why is not the list of failures from small-pox brought forth? My friend, John Ring, had this in progress some years ago; but nothing appears in a compact form from any quarter. No less than seventeen of such cases have been found in the families of the nobility. The late Mr. Bromfield, whom you must recollect was surgeon to the Queen, abandoned the practice of inoculation in consequence of his failures, one of which was at the palace, from an inoculation with a portion of the same thread as was used on the arms of the Duke of Clarence and Prince Ernest, the Queen's brother. Is not this a precious anecdote for your new work?

The above was written long before the expiration of the last year. I have just copied it, as the paper was injured, and you must take it as a proof that I do not intentionally neglect you. It gives me pleasure to think that your second volume is so nearly completed; but I pray you not to let it go before the public eye till it has passed the ordeal of mine. Many new lights have been let in on the vaccine practice most certainly since my own observations first appeared. With regard to the late-formed matter and the scab, there is still a field open for further experiments. I will communicate one to you that I made not long since. Several punctures were made in the arms of a healthy child with vaccine matter, taken from the edges of the vesicle when three-fourths of the centre were incrusted. Not one of them took effect. Some weeks afterwards, with a solution of the same scab, I vaccinated effectually. This, I think, may be accounted for— the scab is made up of the *early* as well as the late-formed matter. On this point, I was certainly cautious in the instructions I first gave out; for an error on my part in this particular could not possibly be injurious to the public. Those who attacked me on this subject made themselves ridiculous, as they made me say what I never said.* I am

Especially with respect to the areola.

grieved to observe that we do not think more alike upon a practical position that I have long laid down. What it can arise from, I am quite at a loss to discover. I allude to diseases of the skin coincident with the progress of the vaccine vesicle. You name to me a case of tinea capitis under which a child was vaccinated, and every thing went right. Why, my friend, that is the very exception* I made in my paper on the subject. It is the minor affections of the skin—what you are sometimes obliged to search for with some diligence,—which more frequently occasion the impediment. If I am deceived in this, every pupil I have ever had in this country is deceived also; for they all remark it.

You tell me you have got a good report for the present year—that is a good thing. The inclosed paper will amuse you, but probably be of no further use. The drawing of the temple which accompanied the paper is mislaid : however, you will find it, I think, in your Pantheon, as it was in that of Hygeia at Rome. What gratitude ! and in a region so distant ! In what part of Britain, should you and I take a ramble, could we discover any thing like this?

I suppose it will be my fate to summer among my oaks and elms (if I summer at all) at Berkeley : but the book— this is a most interesting thing. Of course you have your proof sheets, or sheets of some description. Could not you send me these in succession, and I will really look them over and send them back with all the care and expedition in my power. Try me once more. I feel as if I should be faithful—this is odd language—but my meaning is, that I shall execute the task with fidelity and despatch, unless physically prevented.

I have a great respect for your elegant cousin, and am happy to hear she is about to form an union in every respect

---

* Not an absolute exception. The contrary appears in the case of Church, published by Willan.

so promising. Give her my best congratulations, and unite with them those of my daughter, to whom you were all so kind when she was in town; and I must not forget to thank you for it. Pray send me a list in your next of the *Board*, of which you speak so handsomely. I know Latham and Norris.

To refer once more to your last letter. You must not risk it as an axiom that the lymph of a regular vesicle, *whenever taken*, will excite only a regular vesicle. It might, indeed it would, in *A*, with a sound skin; but it is ten to one if it would in *B*, with a skin on which any of the irritative eruptions appeared. The disturbance of the specific action going forward in the vesicles by the rude thrust of a lancet, is what I have often named as more likely to weaken or even destroy its power than robbing it of its contents by means of a delicate touch or touches. The following is a curious fact; its proof occurred in a village in this country not long since. A female had been vaccinated by means of a single puncture—a good vesicle appeared—from this several of her neighbours were vaccinated at different times during its progress. The woman caught the small-pox, and had it severely. This excited alarm among those who had received the infection from her; they were all subjected to variolous inoculation, and all resisted it.

I have told you ere now that I dislike the appearance of a large, irregular cicatrix after vaccination as much as I do one that is but just perceptible. A young lady whom I lately chanced to see, and who had been vaccinated when a child, had a mark of this description on her arm, and one only. I mentioned my suspicions, and she readily allowed me to insert some vaccine lymph. The consequence was, the appearance of five vesicles, which passed through their stages correctly, and from which I vaccinated with perfect effect. I have inclosed a *proof impression*, the seventeenth in succession, and hope you will greatly admire my ingenuity, as well as the amazing length of my letter; but lest you should have too much of a good thing, I will conclude.

Believe me, dear Moore,

With best affections, most truly yours,

EDWARD JENNER.

I am half disposed to think the first part of my letter was sent to you before; but, right or wrong, it must go now. You see how muddy my head is.

Did you ever see my communications to Willan, published in his work on the Vaccine? You can easily get it.

Berkeley, 10th March, 1817.

P.S. On second thoughts I shall send your letter, intended for the post *direct,* to the care of Dr. Hervey, as I believe it will reach you on the committee day. You will be able, I hope, to give a good account of your Essex expedition. Pray call to your recollection the inquiry that formerly took place at Ringwood. The patients of one medical man there, were almost all susceptible of small-pox after supposed security from vaccination, while those of another escaped the contagion. The true mode of conducting this process is, for the most part, very imperfectly understood every where.

The inconsiderate have a shield. If they fail, no blame, they would have their neighbours believe, attaches to them; it is the thing itself that is imperfect. They know no more than the mere outline of the practice, that is, taking the lymph from one arm and inserting it into another. Sometimes, indeed, they err with their eyes open. A medical man, not far from hence, a short time since was called upon to vaccinate a number of paupers for a certain sum. The only source of infection was two very imperfect vesicles, or rather *pustules,* on the arms of a scabby child, which I had condemned as deceptious. Notwithstanding this sentence (being in the situation of Shakspeare's apothecary), the job was done, and he brought home beef and mutton in his pocket.

I shall send up in a few days some *newly created* vaccine lymph.

Ever truly yours,
EDWARD JENNER.

The following letters being of a miscellaneous nature, and not connected with vaccination, have been placed in their present order, that they might not interfere with that subject.

To THE REV. DR. WORTHINGTON, SOUTHEND.

Berkeley, 25th Sept. 1809.

MY DEAR SIR,

Before I say any thing of your second letter, allow me to thank you most sincerely for your first. You endeavour to cheer me, and that is very friendly and kind of you. Mrs. Jenner as well as myself is sensible of your goodness, and begs her best thanks. There is no material alteration in the state of poor Edward since I last wrote to you—no return of hæmorrhage.

The epidemic you speak of has not been observed here; at least it has not come under my observation in any degree beyond the common run of these maladies. Your mind, I know, never sleeps over human calamities. You speak of prevailing ophthalmia. Excuse my suggesting the great benefit you may bestow on your neighbours by the free use of the unguent. hydrar. nitrati. Theory says, " use it only when the eye-lids are affected; " but practice says, " spare it not when the eye itself is as red as a cherry." In short, I have been in the habit of using it in ophthalmia, under all its varieties, with the most decided success. In cases of the most violent kind, and which quickly threaten to destroy the eye, I introduce a seton in the temple, about an inch from the outward angle of the eye. The latter practice has, I really believe, given sight to thousands since I first made it public, about the year 1783. I now make you my debtor, by giving two *receipts* for one. I shall put about your plan for making good butter, but Prejudice is a giant: however, I shall fling my *pebbles* at him as hard as I can. Your experiments seem to have decided the superior excellence of the horse-hoe, and I hope you will give a paper on the subject to the Agricultural Society, of which I have the honour to be a member, and should be proud to transmit it.

I should have been happy in seeing your nephew as he passed along. Newport is only a mile from my residence.

You must be disappointed at finding a certain vacuum in my letter—no vaccine matter. The fact is, I have none at present but what I fear is unfit for your use. Such numbers have been vaccinated around me, that I have worked myself out of employ, and can now only catch a subject occasionally as it drops into the world. I shall have one soon, when you shall be immediately supplied. But if you are in a hurry pray write to the National Vaccine Establishment in London. They like to have applications from professional gentlemen in the country. Direct as follows,—Dr. Hervey, National Vaccine Establishment, Leicester Square, under cover directed to the Secretary of State for the Home Department.

Believe me, my dear Sir, most truly yours,

EDW. JENNER.

TO THE REV. DR. WORTHINGTON.

Berkeley, 13th Dec. 1809.

MY DEAR SIR,

I certainly have delayed answering your letter of the tenth of November beyond any reasonable, and I almost fear, pardonable period; but if you can forgive me, pray do. Nothing would plead my excuse so forcibly as your seeing the confusion in which I am doomed to live, and nothing but your seeing it would give an adequate idea

of it, for it defies the power of description. I am
by accident, you know, become a public character; and
having the worst head for arrangement that ever was placed
on a man's shoulders, I really think myself the most unfit
for it. You may form some judgment of my accumulated
vexations, when I tell you, that I am at this moment more
than a hundred letters behindhand with my correspondents.
I have lately been deprived of the aid of my secretary.
He was cut off by the same dreadful disease, which, I fear,
will shortly take from me my son. He, poor fellow, still
exists, though I cannot but consider his case as hopeless.
His cough has somewhat subsided, but his pulse is seldom
under one hundred and twenty, and he is extremely ema-
ciated. One thing is remarkable, from the commencement
of the disease to the present period, I do not think the se-
cretions discharged from the lungs (pus, mucus, or whatever they may be) would amount to half a pound in weight.
Allow me, my dear Sir, to thank you for the very kind and
soothing manner in which you speak of him. Mrs. Jenner
feels this as well as myself, and desires to join her thanks
with mine. What dreadful strides pulmonary consumption
seems to be making over every part of our island. I trust
some advantage may, one day or another, be derived from
my having demonstrably made out that what *is* tubercle in
the lungs *has been* hydatid. But I must not tell you a long
story on this subject now, as you must be impatient for my
going into another. I was quite delighted with the detail of
your successful experiments in the profitable science of agriculture, and am happy to find you have finished your Report; but if my letter is not destroyed, and you can refer to
it, I believe you will find that I told you I was a member of
the Board of Agriculture, meaning that in London. If I
said the Agricultural Society, you have certainly been led
into an error. Thinking your observations worthy of the *first*
society in Europe, I did not look to the *second. Utrum
harum?* It may not be material. Your design, I know, is

to impart knowledge; and if your paper is drawn up for the express purpose of going to the Bath Society, I can convey it there with great ease, being intimately acquainted with the presidents of both. To convince you how attentive I have been to your letter, you must know that I made an effort to have a drop of good cider in my house as well as yourself, and imprisoned, as firmly as I was able, a hogshead of apple-juice fresh from the mill; but about the tenth day it seemed so determined to break loose, that to prevent the bursting of the cask, I was obliged to give it liberty: perhaps I was not sufficiently expeditious, for it must be confessed that one half of it had been exposed to the air the day before the whole was bunged up. I anticipate great crops of potatoes, &c. &c. if I live to see another summer. Pray do not suffer what I have said respecting my pile of letters to deter you from writing to me, if you can put up with such a bad correspondent as I have proved myself to be. Indeed, if you do not write to me soon, I shall think you are offended, and believe me that would make a heavy addition to my burden of cares.

With great regard, my dear Sir, truly yours,

EDWARD JENNER.

TO THE REV. DR. WORTHINGTON.

MY DEAR SIR,

I know you will require no apology from me for suffering your last letter to remain so long unanswered. You know the sad movements of the mind in a case like mine, and how it sits brooding over melancholy, unless absolutely dragged out of it. I have placed your letter in my view for some time past; and it has at length urged me to take my pen and answer it. First, let me thank you for your little essay on consolation; you are perfectly right; a person under affliction had better be left to his undisturbed meditations. But, my good Sir, you have been useful to us without being conscious of it; you have inculcated that great Christian principle humility in so impressive a way in one of your sermons, that I feel greatly obliged to you for it.

After the account I have given of myself, you may suppose your Agricultural Report is still lying among my papers. Believe me it has long been a hundred miles off, and in the hands of Lord Somerville. Indeed, ere now I should suppose it has reached its place of destination, the Board of Agriculture, where I anticipate its meeting with the reception it merits.

The state of your drilled wheat, I hope will make converts of the surrounding peasantry. The difference between yours and theirs is this : yours had a plentiful larder to go to, while theirs was starved, or at least had not sufficient supplies to keep out the cold. It was that dreadful frosty night which came suddenly about nine weeks ago, that made such havoc among vegetation. Its effects are every where visible here. Pray do not part with your *free martin ;* it will be a beautiful animal, and docile and useful in your fields as the ox. I have dissected many ; but why this mingling of the sexes should arise under such circumstances, eludes all my guesses. Some of the tricks going forward among the inhabitants of the uterus I have long since pretty well made out ; but this is too much for me. I was the first who made the fact known (some thirty years ago) to Mr. Hunter. He soon went to work upon the subject, and the result was an excellent paper in the Philosophical Transactions. It was re-published in his work on the Animal Economy.

I want to have a deal of talk with you on matters of this sort before I go hence. Oh that you had but taken the Pedington farm ! But it is wrong to repine, all is right. We see through a mist, and shall till our eyes receive a new lustre. God bless you and yours !

Yours, my dear Sir, most faithfully,

EDWARD JENNER.

Berkeley, 5th April 1810.

P. S. Your son has my best wishes. From what I have heard of Mr. Freer, his situation must be very promising.

To the Rev. Dr. Worthington.

My dear Sir,

I received the inclosed by last night's post, and hasten to lay it before you, both for your credit and my own too. I am unfortunately a little given to procrastination; and my character, I have reason to apprehend, is beginning to be known some *thirty miles* north of me. But a word in extenuation—I am more apt to neglect my own affairs than those of my friends. This mode of conduct *we* philosophers can account for. A man will get censured for neglecting the latter; but with regard to himself, he can easily accommodate the matter. Having proved, then, by the inclosed *certificate* from my Lord Somerville, how well I have executed the business entrusted to my care, I hope it will recommend me so strongly to your attention, that you will take me into your service whenever you can make me in the least degree useful.

I suppose you are now in the midst of that pleasant branch of agriculture, potatoe-planting. What a gift from Heaven was this extraordinary vegetable—a ready-made loaf—and reserved too, till the hour when population, in these realms at least, began first to increase; and then coming we scarcely know how. Away with Malthus and his dreary speculations! The skies are filled with Benevolence, and let population increase how it may, let us not distrust it, and suppose that men will ever pick the bones of each other. To descend a step or two, I find that at this season of the year and at the first coming of the potatoe, in order to have it in perfection at my table, I must deviate from that mode of cookery (certainly the best from October to March), which has been pointed out by the Irish, namely, putting it into cold water and suffering the water gradually to rise to the boiling heat; but *now* a plunge into boiling water at once is the thing, where it must remain till the process is finished. Steam I found equally bad with the *cool regimen;* it renders a potatoe viscid and watery, which dressed in the other way is mealy, and readily crumbles under the knife or spoon. If any thing further can add to the improvement, it is a little steaming, when the net is taken from the pot, under close cover. Will your cook pardon this impertinent intrusion upon her province ?

Believe me truly yours,
EDWARD JENNER.

Berkeley, 25th April, 1810.

From Lord Somerville to Dr. Jenner.

Pardon me, my dear Doctor, for not replying sooner to your obliging letter; but I have had a cold-street, hot-room, silk-stocking, champagne fever in London, which has confined me to the bed for some days, or you should have heard from me before.

When my great sale of sheep is over, which will be in eight days, I shall have both leisure to present and pleasure in doing the needful with Dr. Worthington's excellent Treatise; it now lies in my portfolio ready for action; but I wish to be there when it is read.

You keep aloof in this new Vaccine National Establishment, and wise you are in doing it, for well I know that the mean spirit which presides sometimes, of jealousy and intrigue, is hostile to your nature; and you are now enabled to keep a whip ready for the backs of those who play foul in it; in this way you will be of twice the use you could otherwise be. In every sense of the word I am alive to every thing that can do you honour or profit. When you come to town you owe me a visit at this farm, which for purity of air and beauty of views can hardly be equalled.

Ever very sincerely,
My dear Sir,
I am yours,
SOMERVILLE.

Fair-Mile Farm, Cobham, Surrey.

To the Rev. Dr. Worthington.

Berkeley, 4th May, 1810.

My dear Sir,

I have been favoured, since my last dispatch to Southend, with your neat little Essay on Vaccination and your observations on *dipping*. Have you seen an account of some bold Vaccine transactions now going forward among the medical men of the county? Their resolutions appear in the Gloucester and Cheltenham papers. Your county I hope will soon follow this laudable example. The small-pox will never be subdued, so long as men can be hired to spread the contagion by inoculation.

With regard to the other subject you mention, be assured my thoughts have not been idle upon it, having lived man and boy much beyond half a century in a dipping country. Pyrton Passage, four miles only from this place, has been noted for this practice time immemorial; and true it is, I never saw or heard of a single case of hydrophobia after dipping in the Severn, or as our friend Westfaling has it, drowning; for so it is, as you shall hear. I once asked a long-experienced professor what length of time he kept his patients under water? His reply was, " As to that I can't tell, but I keep them under till they have done kicking, when I bring them up to recover their senses and get a little breath, and then down with them again, and so on to a third time, observing the same rule, not to take them up till their struggle is over."

You see then what a shock the vital principle receives from this process. The modus operandi let us not trouble our heads about, if the fact can be established that it deadens the action of the inserted virus. I have wished to see how far it can be supported by analogy, by getting some vaccinated patient dipped within a few days after the insertion of the vaccine lymph. At all events an inquiry so highly important should be taken up, and it cannot be in better hands than yours.

The case of the unfortunate farmer is extremely interesting, and I look forward to your reports upon it with much anxiety. A person is certainly in more danger after receiving the poison on the hand or the face than on other parts, for obvious reasons. The tooth must be wiped by the clothes before it can reach them.

I expect to hear from my Lord Somerville as soon as your papers have been presented and read.

Believe me most truly yours,

EDWARD JENNER.

TO THE REV. DR. WORTHINGTON.

London, Fladong's Hotel, Oxford Street, June 26th, 1811.

MY DEAR SIR,

A great bundle of letters has just reached me by the Gloucester coach. Yours of the 20th is among the number, and if I do not, in spite of the worries of this shocking place, take immediate notice of it, what will you think of me? I am much obliged to the bed of nettles; they have introduced you to a very pleasant family, with whose ancestry I have heretofore played all the pranks you speak of. I have gone further, and entertained a tea-party by placing the young cuckoo, when about four days old, on the table, in its little twiggy cottage, where I have caused it to exhibit its wonderful performances of discarding any thing placed there not too ponderous for it to carry up to the edge of the nest and throw out. Pray be attentive to your young charge, as you will be able to confirm what I have said on this extraordinary subject. A little search may perhaps bring more nests to your view.

I told Westfaling, in a conversation on dipping, that there might be bad dippers as well as bad vaccinators, for which there seems at present to be no allowance. Pray do not be deterred from prosecuting your inquiry. Yesterday I dined with Professor Davy. I wish you had been with us. His mind is all in a blaze. He seems to be one of those rare productions which nature allows us to see once in a score of centuries. We touched on hydrophobia. He started an ingenious idea, that of counteracting the effects of one morbid poison with another. What think you of a viper? Not its broth, but its fang, as soon as the first symptom of disease appears from *canination*. If this should succeed, we must domiciliate vipers as we have leeches. But from this hint I should be disposed to try, under such an event, vaccination; as it can almost always be made to act quickly on the system, whether a person has previously felt its influence or not, or that of the small-pox.

An answer to one of your questions. I am sure the cuckoo has nothing to do with hatching, as all the adults *are off*, while a great number of their eggs remain unhatched. I should put dogs quite out of the question in the new research, and confine myself totally to the human animal; I mean, with respect to dipping.

Success to your crops. I should like to see them before they fall beneath the sickle; and do not yet quite despair. My stay here will be a few days longer only. You can never write too often, or too much to me; but how can you put up with such shocking returns for your kindness? When your cuckoo has gone through all his manoeuvres, pray give me your notes.

Believe me,

Most truly yours,

E. JENNER.

To Mr. E. GARDNER, FRAMPTON.

Gloucester, Saturday, April 13*th*, 1816.

DEAR GARDNER,

I do not think you have written to me since the time you

promised to spend the Easter vacation at Berkeley. It
would be a shock to you to stalk into the old cottage, and
find nothing within it but chairs without associates, grates
without fires, and, worse still, tables with nothing on them
but their varnish. In good truth I am still at Gloucester,
under the roof of my friend Baron, and have been detained
here the whole of this tremendous assize. My intention is
to quit this place (rendered dreary by the tragic scene at
this instant about to be acted on the horrid platform) to-
morrow, and go to Berkeley; but what renders my return
home a little uncertain is a bad catarrh, accompanied with
sore throat and head ache. If Monday, then, was the day
you fixed upon for coming to Berkeley, pray do not put it off;
my motive for writing being nothing more than taking off
the fear that you might possibly go to Berkeley and be disappointed, and, indeed, more than disappointed, for you
might feel hurt at being neglected by an old friend. I
should like for you to collect the feelings of the country respecting the execution, as I must go deeply into the consideration of the case when we meet. They certainly did not
go out with intent to commit murder. But it is somehow
expected that the meanest individual in the state is to be
acquainted with our penal laws and all their intricacies. But,
in my opinion, this is unreasonable, for no general provision
is made for engrafting this knowledge on the mind. An
outline might be imparted by our clergy, by reading to their
congregations four times a year a sketch of these laws; at
the same time they might be blended with moral instruction:
so that the laws and the evil consequences of breaking them
might be committed to memory at the same time. In short,
the village peasant knows no more at present of the laws
which are to act as restraints on his vicious inclinations—that
is, when they move into paths of intricacy—than the village
doctor does of those of the animal economy. We want a
new school Experience has shewn that the present system
of tuition with respect to instructing children in the know-

ledge of the Creator is faulty in the extreme, and I have
every reason to think, that the plan I have long proposed,
and with which you are acquainted, if acted upon, would
prove of incalculable importance to the rising generation.

Dear Gardner,

Truly yours,

E. JENNER.

P.S. I have every expectation of going home to-morrow.

TO THE REV. W. DAVIES, ROCKHAMPTON.

DEAR WILLIAM,

I must wait patiently till I find an opportunity of
going with you for the purpose of exploring the *Breccia*
rocks in the neighbourhood of Thornbury. For this purpose I must have a clear day and a clear head, that is, I
must seize on some lucky hour, if I can find it, when I can
get rid of perplexity, and think of one thing at a time. If
one could manage *the rays of thought* as easily as those of
the sun, bring them to a focus, and to bear upon a particular point, what clever fellows we should be! Perhaps too
clever for the scheme of Providence: so we must take things
as they are, be humble, and be thankful.

Your brother Edward, I find, is very soon going to town;
will you desire him to purchase for me, at some reputable
shop, a few packets of good garden seeds, such as carrot,
onion, lettuce, &c. &c. and the most dwarfish of all the dwarf
peas? There is a sort which grows scarcely higher than this
sheet of paper, and are excellent bearers. I am going to
Kingscote to-morrow, to see poor little Caroline. I fear,
poor thing, the injury will prove too severe for her, and that
she will sink from the extent of it. What pity it is, that
precaution with regard to fire is scarcely ever attended to in
our nurseries. A shower bath, constantly *charged*, should
be ever ready as an instantaneous extinguisher. This I have

been recommending very generally for many, many years, but I never heard that one was put up in consequence. Remote evil is seldom heeded.

Adieu! Affectionately yours,

EDWARD JENNER.

Berkeley, Jan. 14, 1818.

To the Rev. Dr. Worthington.

Berkeley, January 26, 1818.

You speak to me, my dear doctor, about indulging hope. I have almost done with this business, and it is very odd one should continue to grasp at it so long, when it is as slippery as a soaped pig's tail. Did you ever watch little boys running after butterflies? A pretty picture of Hope this. And now about *corporal strength* and animal spirits.

The corporal is in tolerably good condition and fit for service; but of the latter, if I give any account at all, it must be such a miserable one, that I will spare the feelings of a friend, and say nothing.

What, poor Maria not well yet? The fashionable remedy is laurel leaves, made limp by the fire like the leaves of the cabbage, when used as an application. As for myself, I have not a fair chance, as I am tossed about in carriages from morning to night over roads I should suppose as bad as ever the coachman of Julius Cæsar drove over. You little think what a condition this Swindon-batter'd shoulder of mine is in—seldom free from pain by day, and at night it often so terrifies poor quiet Morpheus, he won't come near me. What is all this about Beavan and the Blues?

I must not forget to tell you that I have a weekly stock of vaccine fluid, some of which shall become solid across the Atlantic whenever you will order it. A letter at the same time might be useful, as the matter (which I shall take care to mention) has not been many months taken from its original source; and all they have now in use in America has been passing there from arm to arm for nearly the fifth part of a century.

Catherine is still on the hills at the ill-fated house of Kingscote, where she officiates as first nurse. I begin to think the burnt girl will recover. Poor dear Harriet's case remains undetermined. I shall never prevail on any one to keep a shower bath in some corner of a nursery, *charged*. Were a child on fire, it might be extinguished in an instant; and, indeed, just as soon on a full grown female. Well; such a letter as this for length has not been thrown off the nib of my pen for many a month. Shall you be ever able to get through it? Certainly not, if I go on much longer; so, adieu, my good doctor, and with kind regards to all at the *Albion*, believe me most sincerely yours,

EDWARD JENNER.

TO THE REV. DR. WORTHINGTON.

MY DEAR DOCTOR,   *Berkeley, May* 2, 1818.

I suppose I am got into a sort of scrape with you, but it will be very strange if the day should not extricate me. Three letters from Swindon! All prime, too—right genuine; and not one answered yet. Too bad! There is my confession; take it, and be merciful.

You must be impatient to know something about my petition to the India House in favour of Mr. Roberts. This has been made some time since, but not the least notice has yet been taken of it. On this I put a construction so far favourable, that it is clearly under consideration. Observe, I did not make my application to the directors point blank; for they are all under obligations to me, and consequently would have thrown my letter with a "pish" under their table, in a moment. It was made to a banker who is intimate with many of them, and who, on a former occasion, got me a cadetship for a young man of this place. I almost envy you when you are talking of the state of your fine vegetables. I can get nothing but a few spring greens. Ragged jacks and jerusalems I will show with any body; but if you want a capital thing, get some *Buda kale-seed*, and sow immediately. The grass, &c. which I put into trenches last summer in my kitchen garden, remains nearly in a state of perfect preservation. How is this? I did it from a rule laid down in the works of the Horticultural Society; which paper, by the way, was copied by the fair Emilia for you. Old John and I, at last, after about thirty years' association, are come asunder; or rather we did separate, and are again forming something like an acquaintance with each other. The *old Celt* dug up all my precious beet-root, just as it was in high perfection, and conveyed it to the dung-heap. Within a week old John felt the loss of the pantry so much, that one half of him evaporated. I am daily expecting packets of seeds from Italy and the south of Spain. Nothing, you know, ripened here last year. When will our seas disgorge the polar ice? If all of it is to make the *tour* of the Atlantic, what will become of us while this is about?

What could destroy poor Griffiths? It could not be fulness in the head. When his successor is established, and all the new arrangements are completed, then for a complimentary reply to a certain paper on the subject of *flannel* in contact with the skin. If ever taraxacum, the tooth of the lion, bit off the head of disease, it must have been at such a season as this. It never appeared to be more sharply set. "Farewell, a long farewell." I fear you have all taken to this place.

To the Rev. Dr. Worthington.

Saturday night, Feb. 13, 1819.

Yes, my good doctor, so it is, and so it ever will be. The laws enacted by that mighty Potentate, whose government will have no end, will never be repealed. The poor dear woman, whose untimely loss we now deplore, had, it seems, for several months past, those premonitory whisperings in her ear, which led her to believe her days were nearly at an end. How often we witness this; and in those, who, like herself, were apparently in good health. My nephew carried his heavy load of affliction with a firm and steady step, and is making some admirable arrangements for the welfare of his young family. Emily, I believe, is acquainted with one, which seems to be highly promising.

Ere now, I trust you are liberated from the dreary charge you took upon yourself in your own family; but when you say you have been long watching OVER the afflicted domestic, do you really speak literally? If so, I must say you have been unnecessarily bold. I saw it carried to an extreme in the *first case.* Watching the progress of this epidemic as I have for several months past, I am warranted in saying that it is more contagious than any thing of the kind ever witnessed by me before, but far less destructive. The fatal arrow seems aimed at the brain; and if we can so blunt its point that it shall not penetrate too deep, we have done every thing. You have got Bateman. He seems better informed on the subject than any author I have met with; but why he should entirely discard antimony from his remedies, I cannot conceive. We have all, perhaps, our prejudices. One thing I cannot help naming to you, as you have been very heedless about it; and that is the use of the anti-pestilential vapour. From a wide range of observation, I can speak positively of its guardian powers. We have at last imported the disease into this place. Henry Jenner, who, though he has seen nearly half a century fly over his head, has not yet begun to *think,* perched himself in the midst of a poor family pent up in a small cottage. It was the abode of wretchedness, had the addition of pestilence been wanting. He was infected, of course; and his recovery is very doubtful. I am told to-day that he is very full of an eruption, the appearance of which stands midway between small-pox and chicken-pox. This has been spoken of by some of the

Dublin and Edinburgh authors. The Cheltenham Chronicle certainly appears here weekly, but I seldom see much more of it than its cover. On searching, I have found your second and third number, but shall defer my critique till I find the first. Why did you not mention your design upon *us* sooner? To say the truth, I begin to lose my relish for the inquiry; and not only this, but all others. Yet why should I discard fossils? they will soon be my associates. Never did I spend so cheerless, so wretched a winter. I am become a " sheer hulk," my masts and rigging all shot away.

Old Nixon was a wonderful fellow; but what this unnatural season is to produce who can tell? Its physical consequences will ere long appear. Nothing, I hope, will happen to destroy a certain vegetable yclepped Nicotiana; if so, the little remnant of my comfort is snatched from my life, and all is lost!

Your whole house have the best wishes of, my dear doctor,

Yours most truly,

EDWARD JENNER.

On looking, I perceive it is your middle paper that is missing.

To THE REV. DR. WORTHINGTON.

Berkeley, Sept. 4, 1819.

MY DEAR DOCTOR,

It was not till within these three days, that I heard you had once more bent your steps towards your Gloucestershire dwelling. Some reports had sent you into France, and others had made you a wanderer among watering-places on our own shores; but I am happy in the *vivâ voce* evidence of a reverend divine, who was your fellow-traveller, in finding that you are again breathing the air of our country. May I hope that ere long you will take a mouthful or two of that which sweeps over the meadows of *our* ever-green valley. Are you alone? I hear nothing of your having companions, save and except poor old Tartar, Minx, and her kitten. *I* am in perfect solitude, and have been so these six weeks. Mr. Fitzhardinge is *grousing* in the Highlands, and Catherine is in Yorkshire.

My hot-house has been beset by a new species of *white blight ;* it differs somewhat from that which has so long beset our apple-trees ; but great has been the havoc it has made among the vines. Know you how to destroy vermin of this description ? One occurrence is worth remarking : the trees at each end of the conservatory, which were exposed to frequent fanning by the opening of the doors, are in the highest vigour, free from vermin, and bearing most luxuriantly. It shews us how necessary ventilation is to vegetable life. As the affair between me and letter-writing is nearly come to a termination, I shall desire Stephen Jenner to make a fill up by throwing one of his sketches into the vacant page.

Most truly yours, my dear doctor,

EDWARD JENNER.

P. S. Stephen, I see, has played *old scratch* with the paper ; but it must go, and you must keep it till we send a better. He has done a country auction, and grouped about thirty figures. In my opinion, it is a production of uncommon merit.

We are all at a loss for a precise direction.

To Dr. BARON, GLOUCESTER.

Berkeley, Monday, Jan. 12, 1821.

MY DEAR BARON,

I am frequently hearing of your amended health, and hope soon to find that you are wound up to your usual standard, and able, without the aid of wheels, to come and spend a day at the Chantry. If you do not come, let me have a line soon.

I cannot get my nerves in good order. Certain sounds, such as I am frequently exposed to, still irritate them like an electric shock. The blunt sounds, such as those issuing from the bells in the tower, two pieces of wood striking each other—indeed, *obtuse* sounds of any kind—do not harm me; but the sharp *clicking* of tea-cups and saucers, tea-spoons, knives and forks on earthen plates, so distract me, that I cannot go into society which has not been disciplined and learnt how to administer to my state of distress. But, my dear Baron, I will not repine, I have enough and enough of mercies to be thankful for; and trust you never will find me ungrateful to the Almighty God who bestowed them. May you have his blessings! Adieu, my dear friend,

With best affections, most truly yours,

EDWARD JENNER.

A letter of mine, which I never expected to have seen in our County Paper, appeared there this morning in a state of perfect nudity. For the benefit of the people in a district in Wiltshire, I allowed some of their chieftains, who earnestly entreated me, to publish it in the Devizes paper; but it was preceded by a letter from the 'squire, and followed by another from Dr. Headly, a man of the first reputation and respectability in the county of Wilts.

To THE REV. DR. WORTHINGTON.

Berkeley, Feb. 24, 1821.

I have written but seldom to you lately, my dear doctor, for I have met with very little worth writing about; at least very little in which you would feel interested. Yet you would have been plagued with a letter or two full of nothingness (excuse the paradox) had not my old ill luck pursued me. I rise in the morning tolerably active, and disposed to work with mind and muscle, and actually do work, though scarcely half an hour in the day in the way I could wish, from incessant thwartings and interruptions. What is to be done, then? Those who understand all this, will not corrugate their faces at me; those who do not, will: but it must be borne.

Another thing, too, must be taken into the account; though I boast of my strength in a morning, yet evening seems to come before its time. My afternoon is all evening, and my evening midnight. Such are the uncontrollable workings of the old partners mind and matter (body and soul, if you will), after the firm has been very long established.

I hear you like Bristol; and that the people behave more than civilly to you—kind and attentive; that you have received civic honours, and I know not what. Is it so? If you say yes, I shall be agreeably surprised at their civility and discernment. My late patient, I trust, has found benefit from the pleasant weather we have had so long, for though the nights have been a little frosty, the days have been delicious from the total absence of currents of air. My best affections to her and her sister.

The practice of an humble submission to our misfortunes, or what we are apt to suppose such, is the best smoother of the rugged roads of life. But what am I about? stepping into a territory that belongs to you and not to me. Pardon my presumption, my dear doctor; and believe me with best affections,

Most truly yours,
EDWARD JENNER.

P. S. Mr. Langharne commonly goes down to Bristol every Wednesday; and I shall keep this till he sets off. His return is generally on a Friday or Saturday morning. I almost forgot to thank you for the snuffers; capital, like the last patent corkscrew; superior to the *ne plus ultra*.

TO THE REV. DR. WORTHINGTON.

Saturday, June 16, 1821.

MY DEAR DOCTOR,

As I cannot apologise for myself for the long neglect of your numerous and intelligent letters, I must request the favour of you to do it for me. Neglect, do I say? I should not bring a false accusation against myself neither. It is the constant occupation of my mind on subjects that imperiously demand attention, that distracts and tears me away from what would be far more pleasant to my feelings. The pull is unequal, and go I must when the tug begins; for the public have hold of one end of the rope, and an individual, only, of the other. You see, then, how the matter stands; and I feel certain that you will plead for me if arraigned at the Inchbrook bar.

You have been kind enough to say a great deal to me on the score of health, and in two points I have profited materially by your monitions, namely, exercise and diet. The mile before breakfast, briskly performed, is a capital prelude to the correct movements of the living machinery for the day, at least this puts all into right tune; it resins the bow, and puts all the pegs and screws in their right places. Two miles before dinner, and a pretty long see-saw walk after, settles the account between me and my *props*, as far as the affair of exercise is concerned; but how stands the management of the interior? Thus:—I indulge the natural demands of the stomach with larger supplies both of wine and animal food, even to the libation of two full glasses of bronti, and sometimes, on gala days, to as large a potation of cider— wine glasses, mind me. The scoop is an utensil I cannot touch without burning my fingers. My sleep is sound, and I enjoy enough of it. I go to bed at eleven, and rise before eight. Once more, and I have done with my egotisms. The clicks still annoy me; but far more faintly than when you were here last. Thomas, as usual, daily plants his batteries; but though he seems to load his artillery to the very muzzles, the balls do not get through the cranium and penetrate the interior. In addition to the remedial history, I should tell you, that commonly more than once a day I have taken a weak solution of carbonate of soda, which the learned among us now insist upon it, is the best and most wholesome alkali of the three.

I wish you a pleasant voyage to the east. Shall you not be in town during the bustle of the king's *crownation?* The broad shadows, which the enemies of vaccination endeavoured to cast over it, are vanishing. Sunshine takes their place. Pray look over the Gloucester Journal for Monday next. I am told, that Mr. Richard Hill intends to say a word or two on the subject to some of the faculty in Wotton. Twenty in that small town already slain by the poisoned arrow of Variola! Is not this too shocking? Can you forbear saying a word or two to these murderous people, after you have seen what comes from the brain of the Right Rev. R. H. Look back and invoke the same genius of inspiration that nestled in your heart when you penned the pathetic appeal to the humanity of Cheltenham. You need not be in a hurry. I fear the mischief is not finished. These death-deeds will go on as long as some of the faculty in Wotton can get a fee for their perpetration.

We will endeavour to keep your white terrier till you return. The animal is promising, but, at present, in rather a shapeless state, which, I understand, is to be modelled into the beautiful, by the hand of Time. I have procured a brace of the Genii of your native isle, and prevailed upon them to be placed at your disposal, among your *Penates.*

Not one word yet for poor Mary; and on a rummage I can scarcely find one. The same monotony that is to be found in every sequestered village in the world dwells here. Among the *locomotive corals* very little variety is to be found. The fixtures shine in all their lustre, and passing downward in the scale, attract by their simplicity. Nature's primitive buildings are all constructed on the same plan; the coral is as perfect as the man, as far as regards the stable part of the building. All that we see is shell, or analogous to it, fluids as well as solids. The vital principle is in the interior of the cabinet, under the lock of the DEITY. When this escapes, at once it falls into dilapidation, and is carried, particle by particle, by agencies visible and invisible, through all the regions of the air, earth, sea, and in the course of time lends its assistance in building new mansions, and in rebuilding, or rather repairing, the old. This for Mary. It is a slice off the same loaf she used to get in " days lang syne " for breakfast at the Chantry, sometimes pretty well baked, sometimes not so ; and this, I fear, has more crust than pith about it.

You must be pretty well tired of me by this time. Adieu, my dear doctor.

Believe me, most truly yours,
EDWARD JENNER.

THE REV. DR. WORTHINGTON.

Berkeley, Aug. 2, 1821.

Want of ability, my dear doctor, and not inclination, has occasioned this seeming neglect of you. Here is the old apology come again, and I fear not for the last time. While you have been enjoying luxuries of all descriptions (among the rest the luxury of woe), I have been a fixture in this joyless spot, and here am likely to remain, till removed in one way or another. Perhaps if there were that extent of communication between soul and soul which may be known hereafter, it will be found that I have said a thousand things to you now in inaudible tones, since last we held converse in the ordinary way; but your ear must be new modelled before you can catch sounds of this description, that is, sounds issuing from the tongue of the mind. It almost makes me tremble to speak of sounds, for I am as susceptible as you ever saw me of those *pointed sounds* emanating from the utensils which spread over our dinner and breakfast tables. The blunt noises, such as issue from a peal of bells, I regard not. I stood at the foot of the tower a short time since, and regarded it no more than the hum of Gray's beetle, which now enchants my garden every evening. The cry of hounds and the halloo of the huntsmen would still be music to me ; but the horrible *click* of a spoon, knife, or fork, falling upon a

plate, gives my brain a kind of death blow. Though I soon
scramble out, I am instantly engulfed as it were in an abyss
of misery. You see, then, that I am almost driven out of
society by this misfortune, if one may be allowed to call
any thing a misfortune which occurs to us during our journey through life.

My feelings tell me that I shall not be able to notice many
things you have communicated to me in your letters, for I
begin to flag. Accept, then, this *patchy* scrap; but ere I
quite conclude, I must say a word or two respecting the
land of St. David. I have a sort of mingled feeling about it,
and so have you and my kind friends M. and E. for (why is
it?) mortals of every description, from the sultan to the shoe-
black, when pleasure enters the brain, cannot seal up the cre-
vice through which pain creeps in at the same time, and *vice
versâ*. Till I made the inquiry a day or two ago, I had no
notion the distance from hence, with a carriage, to Abergavenny was fifty miles; but never mind that, when you are
settled, should I be able, I think it highly probable you will
see me there. I shall find no acquaintance but you and your
household. An M. D. lived there not many years since,
who came from Thetford in Norfolk. He was first at Cheltenham, and there I knew him. You must understand by
this, that I am not acquainted with any body at Abergavenny,
unless this gentleman be still there.

Adieu, my dear doctor; with best wishes to my friends on
the banks of the Inch, believe me truly yours,

EDWARD JENNER.

To MISSES M. AND E. WORTHINGTON, CEFN COT-
TAGE, ABERGAVENNY.

Confessions to young ladies from a *young fellow,* such as I am,
are no uncommon occurrences. Speak, Maria, and you, Miss
Emily, are they? You shall hear. What a blundering piece
of work, then, did I make when my last dispatch was sent off

to the Cefn. Instead of returning Dr. Baillie's letter, as I intended, lo! an epistle of the wandering doctor's (Pa's) was sent in its stead; at least, I think so. I could not be quite at ease, touching this blockheady business, till you received an explanation; but mind, both of you, it is not very likely my blunderings will stop here. The *commander-in-chief*, the director of all the forces combined, moral and physical, which form this vital machine, is himself disorderly. Marvel not, then, at deranged emanations, and in future expect no apologies.

The chicken; you shall have some to a certainty, and just those things you describe, made out of grain that passed through my own, my very own, hands.

I am still in solitude here: Catherine at Bath, and Robert lives at the castle. If it were not for a job or two I have promised to perform, I would cut the *black cable* that holds me here, weigh anchor, and sail at once for *Goatland* *. I could bring the poultry with me. What eggs you will have for a spring breakfast, if the doctor does not give a little check to their progress, by chopping the chicken's heads off for disturbing the regularity of his drill horticulture. Have you preserved some chrysanthemums for me? As soon as any of the party can convey some information to me respecting the outlandish country you are got into, I hope they will. What do you call the cottage? Is the C in the Welsh language pronounced like the K?

Well, believe me, I as much intend paying a visit to the cottage, the sweet cottage, sprinkled with the dews (pretty!) from the sugar loaf, as I do to sign my name to this letter, and tell how truly and sincerely I am yours,

EDWARD JENNER.

Berkeley, 15th Sept. 1821.

In your next, say how many letters have been received from me at Cefn Cottage.

* Wales.

To his niece Miss Emily Kingscote, now Lady
Kennaway.

Chantry Cottage, Oct. 16, 1822.

My dear Emily,

More rabbits from Kingscote! So your mamma is not in
dudgeon with me, that is certain; but I should be out of
humour with myself, if for an hour I had *mentally* neglected
her; nay, I have a great deal of intercourse with her. For
I see what I hear, and all the accounts that reach me look
as pleasant as I can reasonably expect. This is not too me-
taphysical for your luminous mind to comprehend.

Your cousin writes cheerfully to her relatives here, Susan
and Caroline. We must all contribute and lend a hand towards the plantation of her flower garden, which she is lay-
ing out most tastefully; and by what I hear, it is nearly
one half the size of your morning room. *" Prodigious!"*

I send this by James Hazen's mother. Her son, at one
period, seemed to be travelling fast towards your church-yard.
I had then the honour to be consulted; and stopped his
journey by the tartar emetic ointment. This is medical, and
must go to your mamma. You see I am thought but little
of in my own household. It must be so; or *unerring lips
would have spoken erringly.*

According to the Almanack, it is a long time to winter;
and I do not despair of coming to see you all before Caroline can give me a snow ball. If I recollect rightly, she is
an adept at this fun; give my love to her, for all that. I like
her mind, at present uncontaminated by *fine ladyism. Excuse word-coining.*

To Dr. Baron, Gloucester.

My dear Doctor,

From the period of our first acquaintance to the present
time, I have been convinced, from a thousand instances, of

your friendly attention; and, I may venture to say, of your
partiality to me. Though your first publication on tuberculous
affections told this tale pretty plainly, yet I am still more highly
gratified at seeing my name prefixed to your last work, under
such high marks of kindness and distinction; and the more
so, as I well know that friendship *only*, powerfully as it ope-
rates on the human mind, would not lead you one inch from
the path of truth and sincerity. This gives a value to your
dedication which I trust I shall know how to prize, and
would, were it possible, rivet my esteem to you still more
closely than before. Having been in possession of your
work but a few days, I have not yet scarcely run over it in a
cursory way, but I like the glance I have taken. My intention is quietly to go through it; and to commit to paper
any remarks, should they occur, for your inspection. May
you long, my dear friend, in the calmness of peace of
mind and health of body, enjoy the fruits of your labours and every earthly blessing. This is the sincere
wish of

Your affectionate and faithful friend,

EDWARD JENNER.

Chantry Cottage, Berkeley, 3rd Dec. 1822.

TO MISS EMILY KINGSCOTE.

Berkeley, 10 *Jan.* 1823.

MY DEAR EMILY,

The carelessness of the carter is a little unlucky, as my letter to your mamma contains a line or two of a private nature,
but not of any great consequence. Sooner or later, I dare say,
it will find the place of its destination. You are very good
in writing so kindly to me, after my *seeming* neglect of you
all. You think me idle, no doubt. Ah! my dear Emily, if
you did but know the laborious work I have to go through,
your opinion would soon be changed. In earlier days, in-

deed at any period of my long life, I do not think there ever was a period when I worked harder. It is no bodily exertion, of course, that I allude to; but it is that which is far more oppressive, the toils of the mind. I am harassed and oppressed beyond any thing you can have a conception of. In the midst of these embarrassments I have not a soul about me who can afford me assistance, except, indeed, my two good-humoured nieces, who copy letters for me, and would willingly do more if they could. When next I climb your icy mountains, do pray see if you and the ingenious inhabitants of the morning-room cannot devise some means to extricate me from my irksome situation. I have a thought:—a silver spoon lies in a small compass; and a voyage to Botany Bay would be a happy exchange for me. Should I have your good wishes, Emily, as I passed over the ocean? I know I should; and you have mine, and your mamma, and all my good friends around you.

EDWARD JENNER.

TO MR. E. GARDNER, FRAMPTON.

Berkeley, Jan. 13, 1823.

DEAR GARDNER,

What a bustle this Frampton watchmaker makes. Your letter is delivered to me whilst I am eating my chop, and an answer demanded immediately; and so, that disappointment in this respect may not add to your catalogue of sufferings, I quit my bone to pack up for you some vaccine matter, fresh and fine, from the arm of Edward Jenner, my young neighbour. With the scab I never fail, nor with the glasses, on which, if you hold them to the light, the inspissated matter becomes visible. Moisten it with a small portion of cold water, and then insert it by three or four punctures, as if it were just taken from the arm. If you use the scab, moisten it with water on the back of a plate, and work it with a little water by means of a clean knife, then insert the matter. If you do not succeed with all this, I shall say you are no pupil of mine, or perhaps call you a bungler, or shall suspect that your patients have eruptions. I begin to fear I shall not see you at Berkeley this Christmas. "Where is Mr. Gardner?" is the cry of my intimate neighbours. I have an attack from a quarter I did not expect, the Edinburgh Review. These people understand literature better than physic; but it will do incalculable mischief. I put it down at 100,000 deaths, at least. Never was I involved in so many perplexities. Metaphysics are on the shelf; but, mind me, I do not conceive there is a single living particle of matter in the universe. The brain, ay, and the nerves, too, are dead as my hat. All life is in that something superadded to matter, *the anima*, diffused through matter, if you will; but to speak like a chemist, not chemically combined with it, not forming an integrant part, but merely influential. "There is something behind the throne greater than the throne itself." Susan and Caroline are at Ebley; Catherine is very well, and I believe very happy. Edward Davies has been on a visit to her, and speaks highly of her situation.

Truly yours,
EDWARD JENNER.

TO MASTER W. DAVIES, HIS GRAND NEPHEW.

This letter was written the day before Dr. Jenner's fatal seizure. It was, I believe, the last he ever penned.

Chantry Cottage, 24 Jan. 1823.

MY DEAR WILLIAM,

I hear by your father that you will return in a few days to Bristol. Be assured you will take with you my best wishes and affections, which I present to you with greater delight than at any former period, because you are more entitled to them, for I am happy to certify, that no boy could behave better than you did during your stay with me at the Chantry. Pursue this line of good conduct, my dear Wil-

liam, and you will be happy yourself, and make your father
and every one who loves you happy too.

Your affectionate uncle,

EDWARD JENNER.

Most of the following compositions are connected
with incidents in Dr. Jenner's personal history. The
first refers to his old gardener John Jones, who was
in his service for nearly thirty years. He died in
1821, and was followed to the grave by his indulgent
master. The second brings before the reader two of
his favourite animals, Minx and Tartar, who were
the constant visitors of the parlour, each occupying
a place on the hearth rug. The third is taken from
a poem of considerable length, entitled Berkeley
Fair, and contains a humorous account of his museum. He introduces himself as the showman, and
carries on the character in a very graphic manner.
Two others arose from little occurrences in the domestic history of his daughter Catherine, and give
pleasing illustrations of the writer's mind. The last
is an enigma full of point, and capable of bearing a
comparison with most similar compositions.

ON SEEING AN OLD MAN MOWING.

Ah, poor old John, with low bent back
See he pursues his steady track;
The wild flowers tumbling at his blade,
Prostrate before him fall, and fade.
Yet little reck'st thou, honest John,
Intent thy 'custom'd work upon,
Of that old mower, grisly Time,
Though long hast thou gone by thy prime;

Witness thy grey locks loosely spread
In lessening numbers o'er thy head.
Thy wither'd cheek, thy tawny brow,
Once smooth and fair, but furrow'd now.
Yes, soon the keen edge of his scythe thou'lt feel:
Look round, old John, 'tis close upon thy heel.

DIALOGUE BETWEEN MINX THE CAT AND TARTAR THE TERRIER DOG.

Tartar. Well, Minxy, you've been out again,
    Killing poor birds for prog;

Minx. And you have kill'd the bantam hen,
    You have, you nasty dog.

Tartar. How very clever and well bred!
    You 've much improved, I see;
    But think on what hangs o'er your head,
    Look up the willow tree.
    Suppose, now, I 'd a mind to tell
    What happ'd within this hour,
    Did I not see thy talons fell
    At work in yonder bower?
    Too plain I heard the dying scream
    Of a poor robin there,
    Too plain I saw the life-blood stream,
    Whilst thou its limbs didst tear.

Minx. Well, Tartar, if you go to that,
    I a sharp word could say;
    Who was it kill'd the farmer's cat?
    Now chew on that, I pray.
    How sheepish now thy looks appear,
    Thou drop'st thy ears and tail,
    As if thou thought'st the halter near—
    Ho! ho! I 've hit the nail.

Tartar. Why talk, pray, of such stuff as this ;
         Why, 'twas but one old cat :
         It cost her but a few short moans—
         Now, Minxy, pray take that.
Minx. Take that, indeed ! Who stole the fish
         The cook miss'd t' other day,
         And spoil'd entirely the dish ?
         Now hold your jaw, I pray.
Tartar. What, not a word ? I see you 're fast,
         Yes, Mrs. Minx, you 're dumb ;
         Well, let us think of what is past,
         And mend for time to come.

Extracts from Berkeley Fair.

It opens thus : —

The sun drove off the twilight gray,
And promised all a cloudless day ;
His yellow beams danced o'er the dews,
And changed to gems their pearly hues.
The song-birds met on every spray,
And sang as if they knew the day ;
The blackbird piped his mellow note,
The goldfinch strain'd his downy throat,
To join the music of the plain
The lark pour'd down no common strain ;
The little wren, too, left her nest,
And, striving, sang her very best ;
The robin wisely kept away,
His song too plaintive for the day—
'Twas Berkeley Fair, and Nature's smile
Spread joy around for many a mile.
The rosy milkmaid quits her pail,
The thresher now puts by his flail ;
His fleecy charge and hazel crook,
By the rude shepherd are forsook ;

The woodman, too, the day to keep,
Leaves Echo undisturb'd in sleep :
Labour is o'er—his rugged chain
Lies rusting on the grassy plain.

* * * * *
* * * * *

Here, neighbours, are sights, such as never before
Were seen at a Fair, and never may more.
Myself and my partner have taken great pains
To display all the wonders of fossil remains.
Now at once, my good friends, you all may inspect
Some remains of the ruins when Nature was wreck'd;
When mountains, vales, oceans, together were hurl'd,
And dread desolation dash'd over the world.
My cabinets all, the subject's terrific,
Shall nothing contain which is not scientific ;—
There's an encrinite's head, a cornu ammonis,
And marquisites fit to adorn an Adonis ;
Fine corals, all fossil, from Woodford's grand rock ;
And granites from Snowdon in many a block ;
Alcyonites, too, we have join'd to our stock ;
Hippopotamus' bones, and the great alligator,
And things most surprising thrown out of a crater ;
All changed into flint are an elephant's jaws,
The mammoth's vast teeth, and the leopard's huge paws ;
There are beautiful agates wash'd up by the fountains,
And crabs that were found on the tops of the moun-
  tains ;
Asbestos, chert, chrysolite, quartz, hæmatites,
Madrepore, schistus, basalt, and pyrites ;
Oolites, zoolites, gryphites a store,
Pentacrinites, chlorites, and many things more.
All this we'll display to those who are willing—
Though the sight's worth a crown—yet for one single
  shilling !

* * * * *

And now the clamours die away,
The sun has sent a farewell ray;
The hills have lost their golden hue,
And wrapp'd themselves in mantle blue;
The showman's voice has lost its tone,
The trumpet's clang becomes a moan;
The Giant now lays down his head,
And Lady Morgan's gone to bed.
The lions all begin to dose,
And tigers seek a soft repose;
The customer no more is courted,
And every standing is deserted:
The Fair is o'er.   But joys like these
Long revel in a heart at ease.
The milkmaid, as she skims her cream,
Long on the happy time will dream;
And many a simple rustic-swain
Will strive to whistle out the strain,
While raking up the new mown hay
All in the merry month of May,
That Jonathan's melodious bow
Bade in his bosom ever glow;
And little girls and little boys
Will for a moment quit their toys,
And cling about their mother's knee,
Asking when Fair again will be;
While every breast with hope will burn,
To see the happy day return.

FROM HER DORMOUSE TO CATHERINE JENNER, 1810.

Start not, fair maid, to see a mouse
Obtrude himself upon your house;
Not one of those sad elves am I
Who pilfer cheese and spoil the pie,

Who through your chambers ever freaking,
Disturb you with a nightly squeaking;
Quite opposite am I in nature
To this intolerable creature.

My birth I boast beneath the bower
Hard by the foot of yonder tower,
Whose battlements o'erhang the place
Long honour'd by a Ducie's race. *
But now I've left my leafy cell
With you, dear Catherine, to dwell.
No dainties seek, I pray, for me,
My food grows on the hazel tree ;
I ask but this, except a sup
Of water from an acorn cup.
Then for a house—oh ! any thing
Will serve for this, that you can bring:
That little box, in which you place
Your pearly trinkets and your lace.
No clothes I want; for, see, I'm drest
By Nature in an ermine vest.
Can looms that weave the satin fine,
Produce a robe so fair as mine ?
Ah ! no—for she will ne'er impart
The means of rivalling her art ;—
Some moss entwined with leaves of willow
Will make an admirable pillow.
You must not take it much amiss
If I'm particular in this,
For well you know I'm one of those
Who like whole months of soft repose ;
And when you come to take a peep
And find your little charge asleep,

* Tortworth, Gloucestershire.

Pity 'twill be indeed to wake me,
And from my dreams of you to take me.
But tho' the hand that made my frame
And my Catherine's were the same,
Another time perhaps I may
Expatiate on what I say;
But now I only can afford
Just time enough to drop a word:
Sometimes I dream that you 're surrounded
By much temptation, and confounded
Just for a moment, when arise
The world's delights before your eyes;
But then, before I end my nap,
I'm sure to see you take the map
Which shews life's road and all its danger
To every inquiring stranger;
The craggy rock, the deep morass,
The precipice, the treacherous pass.
And charm'd am I to see you steer
By the just compass of Montier; *
Which ne'er will lead my Kate astray
From Truth's undeviating way.
Good bye—I know we shall agree,
You will be ever kind to me;
And this shall be my constant plan,
To please in every thing I can.

To a Tom Tit

*who was fed every morning at the bed-room window of
Catherine Jenner, at Cheltenham.*

Oh! tell me why, my dearest Thomas,
You stay'd so long this morning from us?

Miss Jenner's governess.

I peep'd at eight, at nine, at ten,
And then I peep'd, and peep'd again.
But oh! my heart! my pretty bird
Was neither to be seen or heard;
Untouch'd the breakfast I had spread—
Nice apple chopp'd, and crumbled bread;
Yes, and the cup I'd early dipp'd
In the clear Chelt remain'd unsipp'd.
Ah! me, said I, some ruffian from me
Has surely torn my darling Tommy—
Some murd'rous hawk, or ravenous kite,
Hides him for ever from my sight.
And, while thus wailing was your Kate,
Methought I saw what sealed your fate,
For to my window, now alas!
Some downy feathers seem'd to pass;
Feathers so beautifully blue,
They could belong to none but you;
But, sweet to tell, my grief, my sadness,
Changed in a moment was to gladness.
The joy I felt I cannot utter,
When I beheld thy charming flutter;
Heard thy sweet voice upon the tree,
And saw thee look, and look for me:
But I must chide thee, dearest bird,
Indeed I must, upon my word.
Well, well, it sha'n't be now—but then,
Tommy, ne'er serve me so again.

ENIGMA.

Through many an age did I sleep quite profound,
Deep hidden from mortals, beneath the cold ground,
As harmless and quiet as if I'd been dead,
Till insulted by rapine, and dragg'd out of bed.

Then, without any crime, by tyrannical power,
Committed was I, under guard, to the Tower;
There stampt upon, cut, yet it gave me no pain,
Though it made an impression that long will remain.
At length I'd the luck from the place to escape,
And now to all ranks dare exhibit my shape.
When first I forth started, I own it with pride,
His Majesty stuck very close to my side;
But, as I grew older, how hard is my case,
The connexion he quits, and scarce shows me his face.
Tho' the great scarcely own me, the pallid-faced poor
With pleasure behold me come out of a door.
How oft may you hear them, in tones very pressing,
Solicit my visiting them as a blessing!
I'm the ficklest fellow, perhaps, in the nation,
For ever am shifting and changing my station!
Nay myself can I change too, without going far,
For a gingerbread watch, or a quid for a tar.
When alone I'm a pauper—a mate for a clown,
Yet join'd to my comrades can purchase a crown:
I'll give a hint more, tho', perhaps, you may laugh—
I'm one perfect whole, yet exactly a half.*

The letters printed in this section of the work, independent of the information which they convey respecting the most remarkable phenomenon in the physical history of man, are worthy of observation, on account of the naturalness and simplicity of the style. There is an exquisite perception of propriety in the manner of expression, and an ease and freedom in the choice of words, which has seldom been exceeded in any similar compositions.

The subjoined meditations are of a very miscella-

* Halfpenny.

neous character, and are selected in order to afford specimens of the varied powers of the writer's mind ; and the tone of reflection in which he delighted to indulge, and with which his note books and his journals abound.

It is possible that the surface of the earth, or more than the mere surface, forming the Berkeley district, may by one of those vast convulsions (which, it is plain, at distant periods threw the globe into the utmost disorder) have been covered with materials brought from very distant regions. The shells and corals found in the rocks at Gibraltar are similar to those found in the rocks at Thornbury, and so are the fragments of some of the stones. The coral, so abundant in the range of many miles, was never probably generated in the spot on which it now reposes: nor any of the families of stones which lie about the surrounding country. All might have been impelled forward at the same period, driven by the mighty, the irresistible torrents of the great deep rushing from their subterraneous prisons. That these corals, such as appear from the immense masses in which they are heaped together in the rocks of Woodford and of Falfield, have been forced from their original position, seems to admit of demonstration from fragments of these identical corals being found among the Thornbury Breccia. This Breccia appears to consist of portions of rocks varying in size from a large block down to miscroscopic atoms; and it appears to be the aggregation of these atoms which forms the cement which binds these fragments together. These fragments do not appear to belong to any rocks in our neighbourhood; but seem to have been mingled together, and driven over an immense space, by one mighty sweep of that powerful element, water. Not only corals and shells are found among this Breccia, but even fragments of basalt itself. This, indeed, does not prove the basalt rocks to have moved forward with the general sweep; as these might have been detached, and mixed with other fragments by passing onwards.

OBSERVATIONS ON THE NIGHT-BLOWING PRIMROSE.

Walking one evening in the early part of July in the garden of a gentleman at the west end of the town, my attention was drawn towards that curious plant the night-blowing primrose, which was growing abundantly and in great perfection in the borders. The petals of this plant, about the setting of the Sun, burst rather suddenly from the calyx in which they are involved during the day, and immediately display themselves in full expansion. In the morning they are puckered up and withered, without a vestige of their beauty remaining. On contemplating this curious subject, one thing struck me as very singular—the apparent waste of that food (the nectarium) which affords nourishment to so many insects, and with which I found the plant plenteously stored. On visiting the spot again an hour after sunset, the subject still occupying my thoughts, the mystery was cleared up, and the scheme of nature most charmingly displayed. I now saw a considerable number of moths of various kinds hovering about every primrose bush, and passing from flower to flower, sucking through the proboscis the nutritious fluid so admirably prepared for them. These moths lie dormant during the day. Should one of them be accidentally disturbed and compelled to take wing, it is at once beset and taken by those birds, which are eagerly looking out for insects for food for their young. Hence the necessity of some provision for their support about the commencement of twilight, when their enemies have retired to rest. It should be remarked, that the petals of many of those plants which afford the nectarium abundantly are open by day only, and are closed at night, as the jessamine, marvel of Peru—a species of geranium. The garden willows are night-blowing plants, and doubtless, in this respect, destined to answer the same purpose in the system of nature. There are doubtless many other plants which afford nectar to the moth during the night besides the night-blowing primrose;

but I bring the above observation forward as illustrative of
the economy of nature in preserving for the support of a
tender insect, what to superficial observation appeared to be
wasted; the nectar afforded by this plant being in its recep-
tacle but for a few hours, and that during the night only.—
April 7th, 1818.

Physicians will probably find one great impediment to the
progress of their art in the revolutions which take place in
the course of diseases.

The human body is a kind of commonwealth; the seat of
government does not seem confined to one director only, but
to several, among which the brain is the chief.

It is not an universal law that diseases of the stomach
affect the head, and that those of the head affect the stomach.
When W. destroyed himself by taking nearly half an ounce
of arsenic, his head was free from pain, and the intellects
clear, during the eleven hours of existence after he received
the poison into his stomach.

The human body is an immense laboratory, divided and
subdivided into ten thousand compartments, the chief of
which is the glandular system.

We are mistaken if we suppose that the stomach is the
grand sufferer among the vital organs of the drunkard. No,
it is the brain. Whoever will consider the phenomena, will
soon be convinced of this; and from these phenomena may
be deduced the crumbling down of the constitution, by the
dilapidation of the vital organs.

I want to do away with the whole stomach pathology at a
sweep, and to place the brain upon the top of the lofty
pedestal allotted to it; to shew it as exercising a complete
sovereignty over every vital action.

Among the vulgar, the great, and the little vulgar, nothing
is recounted or recorded of vaccination but its imperfections;
its benefits are passed by, and are often forgotten. Vacci-

nation has to contend with more ignorance than any other process the faculty are engaged to attend to ; and in some instances for this reason :—if a failure take place, the vaccinist is shielded, by declaring that it is vaccination itself that is imperfect, and no fault in his conducting the process. If vaccination ever fails, it must be owing to some peculiarity in the constitution. From the year 1762 to 1792, the number that died of small-pox in the Danish dominions amounted to 9,728. About the year 1802 vaccination was first introduced, and the practice became general, but not universal; however, fifty-eight persons only died of the small-pox to the year 1810. Vaccination, by command of the King, was now universally adopted, and small-pox inoculation prohibited. And from the year 1810 to the year 1819, not a single case of small-pox has occurred.

One interesting trait in vaccination is, what every observer of its progress must have witnessed, namely, that every thing has worked together for its good. The opponents have been greater instruments in facilitating its progress than its promoters, by calling up inquiry, which has always ended in full proofs of its efficacy.

Vaccination has been extensively practised in this country, yet very imperfectly understood.

What instance has occurred to throw a real slur on vaccination? I pledge myself to this point. I could not possibly go farther, than that in every instance it would afford security equal to the small-pox inoculation, if the rules I laid down were strictly observed.

The Sacred Scriptures form the only pillow on which the soul can find repose and refreshment.

The power and mercy of Providence is sublimely and awfully displayed in lightning and in tempest. It scarcely ever happens during what is called a thunderstorm but we hear of some human being suddenly losing his life by a flash from the Heavens. And when the tempest roars around us, we know that some destruction always follows. But how beautifully is power here seasoned with mercy! We are shewn, that instead of partial chastisements, it might have been universal. The Almighty arm that struck prostrate a single individual might at the same time have hurled his bolts on the heads of all. He that directed the storm to shew his mighty strength by partial destruction, shews every beholding eye, that at his fiat it might have swept off every living thing. But how beautifully is it modified! it goes just to the point, where every thing terrestrial seems upon the verge of universal wreck, and then mercifully softens into a calm. How sublime, how awful is this display of the power and mercy of God!

Our ordinary language shews us, as it were, unconsciously, our ideas of a compound existence, the subserviency of the body to the agency of the soul:—" I tore myself out of the house;" " I was out of my mind," i. e. my mind was out of me.

" I have heard of thee by the hearing of the ear, but now mine eye seeth thee." The above is applicable to the subject of reform in the education of children. The great book of the world is open to all eyes. My wish is, that every human being might be taught to read it. The poor man does not know what a rich library he is in possession of; that he has an equal right with the proudest monarch on the globe to have access to it. A sincere acquiescence in the dispensations of Providence will check discomposure of mind beyond any thing. It will produce a calm in the midst of a storm.

If we fear all things that are possible, we live without any bounds to our misery.

The highest powers in our nature are our sense of moral excellence, the principle of reason and reflection, benevolence to our fellow creatures, and our love of the Divine Being.

APPENDIX.

No. I.

Chronological List of Diplomas, Honours, Addresses, and various Communications from Public Bodies and distinguished Individuals to Dr. Jenner, on his Discovery of Vaccination.

1801. PLYMOUTH DOCK. *Feb.* 20.—Address from Dr. Trotter, and forty-four medical officers of the Navy, subcribers to the Jennerian Medal.

NOVARA. *May* 29.—Address of respect and application for imbued threads, from the " Physician delegated " of the department de l'Agogna, (Cisalpine Republic.) Signed, Gautieri; Mantillari, Secretary. (In English).

PARIS. 16 *Thermidor.*—Address from the Bureau of the National Institute of France; and thanks for the dissertation communicated to them. Signed, Coulomb, Pt. G. Cuvier, Sre. Delambre.

GOTTINGEN. *Sept.* 14.—Diploma of Fellow of the Royal Society of Sciences at Gottingen. Signed, Henricus Augustus Wrisberg, Philosoph. et Medic. Doct. Britann. Regi a Consil. aulæ. Medicinæ et Anatom. Professor. publ ordinar. necnon Societat. reg. Gotting. h. t. Pro Director.

1802. MANCHESTER INFIRMARY. *Feb.*—Certificate of the

success of Vaccine Inoculation, and complimentary Address thereupon. Signatures. Physicians : Thomas Percival, M.D. Physician Extraordinary, John Ferriar, M.D., Samuel Argent Barclay, M.D., James Jackson, M.D., Edward Holme, M.D. —Surgeons : Wm. Simmons, John Bill, Alexander Taylor, M.D., R. W. Killer, M. Ward, G. Hamilton, J. Hutchinson, House Surgeon; Thomas Henry, John Boutflower, Visiting Apothecaries.

LONDON. *Feb.* 20.—Diploma of Fellow of the Physical Society of Guy's Hospital. Signed, Joannes Haighton, M.D., Thomas Walshman, M.D., Jacobus Curry, M.D., Ricardus Saumarez, Astley Paston Cooper, Thomas Hardy.

Feb. 25 —Testimonial and Address from the Presidents and Members of the above Society. Signed by the six Presidents and one hundred and six Members.

EDINBURGH. *March* 7.—Diploma of Fellow of the Royal Medical Society of Edinburgh. Signed by four Presidents and twenty-five Fellows.

PARIS. 24 *Ventose.*—Diploma of Foreign Associate of the Medical Society of Paris. Signed, Thouret, President; Alibert, Secretaire-general.

TOURS. 30 *Germinal.* — Official Address from the Medical Society of Indre et Loire. Signed, Bouviat, D.M.M. Secretaire-general de la Societé Medicale seante à Tours.

MASSACHUSETTS. *May* 25.—Diploma of Fellow of the American Society of Arts and Sciences in Massachusetts. Signed, John Adams, President; Joseph Willard, Vice-President; Attest. John Davis, John Quincy Adams, Secretaries.

PARIS. *July* 29.—Official Letter of respect and congratulation upon the general success of Vaccination in France, from the Central Committee of Vaccination. Signed, Thouret, Directeur de l'Ecole de Medicine, President; Husson, Secretaire. Conveyed by Citizens Huzard and Parmentier; and accompanied by a Letter from the Secretary of the said Committee.

POWLOWSK. *August* 10.—Letter from the Dowager Empress

of Russia, signed "Marie," and accompanied by a ring set in diamonds.

Tours. 2 *Messidor.*—Diploma of Corresponding Associate of the Medical Society of Tours. Signed, Bruneau, President; Bouriat, D. M. M. Secretaire-general.

Avignon. 27 *Brumaire.*—Appointment of Associate from a Society at Avignon. Signed, Fortia, Vice-President; Hyacinthe Morel, Sec.

1803. *March* 16.—Diploma of Member of the Society of Medicine at Avignon. Signed, Voulonne, Med. President; G. Guerin, Secretaire; Clement, fils, Sec. Adj.

London. *August* 11.—Freedom of the City of London, presented in a gold box of the value of one hundred guineas.

Madrid. *August* 15.—Diploma of Fellow of the Royal Medical and Economical Society of Madrid. Signed, Antonius Franseri, Pro-præses; Hippolytus Ruiz, Rei Pharmac. Censor; Liz. Philipus Somova, Chirurgiæ Censor; Joannes Peñalva, Scientiar. natural. Censor; Casimirus Ortega, in rebus ad exteros spectantibus a secretis; Ignatius Maria Ruiz .... a secretis.

Massachusetts. *August* 31 —Diploma of LL.D. from the Senate of Harvardian Cambridge University, Massachusetts. Signed, Josephus Willard, S. T. D., LL.D., Præses; Oliverius Wendell, Simeon Howard, S. T. D , Johannes Lathrop, S. T. D., Eliphalet Pearson, LL.D., Johannes Davis, Socii; Ebenezer Storer, Thesaurarius.

London. *September* 14.—Diploma of Honorary Member of the Royal Humane Society of London. Signed, President, Stamford and Warrington; Treasurer, John Coakley Lettsom, LL. D. M. D.

Paris. 28 *Vendemiaire.*—Diploma of Foreign Associate from the School of Medicine at Paris. Signed, Chaptal, President; Le Clerc, Secretaire.

Nismes. 21 *Frimarie.*—Diploma from the Society of Medicine, Departement du Gard. Signed, Vitalis, President; J. B. Dubois, Presid honor.

1804. DUBLIN. *March.*—Freedom of the City of Dublin.

PHILADELPHIA. *April* 7.—Diploma of Member of the American Philosophical Society at Philadelphia. Signed, Th. Jefferson, President; C. Wistar, Jun., R. Patterson, Vice-Presidents; Benjamin Smith Barton. Attested, John Redman Coxe, Thomas C. James, Adam Seybert, Thomas H. Hewson, Secretaries.

EDINBURGH *October* 31—Freedom of the City of Edinburgh, transmitted by Sir William Fettes, Bart. of Wamphray, Lord Provost; William Coulter, Archibald Campbell, John Turnbull and James Goldie, Esquires, Baillies; John Muir, Esq. Dean of Guild; Peter Hill, Esq. Treasurer; and the remanent Members of the Council.

WILNA. 16 *Kal. Dec.*—Diploma of Fellow of the Imperial University of Wilna, issued by command of the Emperor of Russia, Alexander I. Signed, Hieronymus Stroynowski, My. Josephus Mickiewicz, Decanus, Professorum ordinis scientiarum Physicarum et Mathemat. Canonicus Cathedralis Samogitiensis, Mpp.; Simon Malewski, Juris Nat. et Gen. Prof. Consiliarius Aul. Secr. Imper. Univers. Vil.

1806. STOCKHOLM. *March* 31.—Diploma of Foreign Associate of the Royal College of Physicians at Stockholm. Signed, Elias Salomons, D. Rung, Joh. L. Odelius, Andreas Sparrman, Joh. Hardtman, C. Von Schulzenheim, L. Hedin, W. fiz. Hadstrom, J'oran Rooth, N. Almroth, Fredric Krey, Loco Secretarii, Conrad Eckerborn.

EDINBURGH. *May* 20.—Diploma of Honorary Fellow of the Royal College of Physicians, Edinburgh.

1807. VALENCIA. *March* 5.—Diploma of Honorary Associate of the Royal Economical Society of Valencia. Signed, Juan Sanchez Cisneros, SSrio. pptuo.

LIVERPOOL. *April* 1.—Freedom of the Borough of Liverpool. Thomas Molyneux, Esq. Mayor.

STOCKHOLM. *April* 23.—Diploma of Foreign Associate of the Royal Academy of Sciences at Stockholm.

FORT GEORGE, UPPER CANADA. *November* 8.—Address of the Five Indian Nations, with a Wampum Belt. See Chapter IV. p. 104 of this Volume.

1808. MUNICH. *March* 28.—Diploma of Fellow of the Royal Academy of Sciences at Munich. Signed, Jacobi, President; Schlichtegroll, Gen. Secr. Moll....

PORTSMOUTH, AMERICA. *May* 25.—Diploma from the President and Fellows of the Newhampshire Medical Society. Signed, L. Spalding, Secretary.

PARIS. *June* 20.—Diploma of Corresponding Member of the National Institute of France, in the class of the Physical and Mathematical Sciences. Signed, Le Secretaire perpetuel, G. Cuvier.

GLASGOW. *September* 1.—Freedom of the City of Glasgow, from the Hon. James Mackenzie, Lord Provost; James Denistown, Nicol Brown, William Glen, John Ballantyne, and George Lyon, Esquires, Baillies; James Black, Esq., Dean of Guild; William Brand, Esq. Deacon Convenor, and the other Members of the Common Council. Jas. Reddie, Town Clerk.

1809. KIRKALDY. *April* 27.—Freedom of the Burgh of Kirkaldy, with thanks for his discovery, from Walter Fergus, Esq. Provost; Robert Brown and James Mackie, Esquires, Baillies; Michael Beveridge, Esq. Dean of Guild; and David Morgan, Esq. Treasurer, and the Common Council of the said Burgh; Will. Drysdale, Clk.

1810. MANCHESTER. *April* 27.—Diploma of Honorary Member of the Literary and Philosophical Society of Manchester. Signed, by the Presidents, Vice-Presidents, and Secretaries; Thomas Henry, Edward Holme, John Dalton, William Henry, B. Gibson, William Johns, I. A. Ransome.

1811. PALAIS DE RAMBOUILLET. *May* 13 and 19.—Diploma of Foreign Associate of the Imperial Institute of France, in the Class of the Physical and Mathematical Sciences, in the room of M. Maskelyne, deceased, and with approbation of His Majesty, the Emperor and King. Le Ministre Secretaire d'Etat, signé, le Comte Daru. Pour ampliation conforme, Le Secretaire perpetuel pour les Sciences mathematiques, Delambre.

1813. OXFORD. *December* 3.—Diploma of Doctor in Medicine of the University of Oxford, from the Chancellor, Masters, and Scholars of that University. Seal inclosed in a gold box.

1814. BORDEAUX. *July* 1.—Diploma of the Royal Society of Medicine at Bordeaux. Signed, Lapeyret, President; G. M. Caillau, D. M. Prof. Royal Secretaire general.

BRÜNN IN MORAVIA. *October* 20.—Address of the Inhabitants of Brünn. Signed, Medicinæ Doctor, Rincolini, Physician Claviger, first Surgeon and Vacciner of Vaccine Institute at Brünn.*

1815. ERLANGEN. *January* 20.—Address of Honorary Associate of the Physico-Medical Society of Erlangen. Signed, Dr. Chr. Fr. Harles, Soc. Director; Dr. Adolph Henke, Secretar.

1821. LONDON. *March* 16.—Appointment of Dr. Edward Jenner to be Physician Extraordinary to the King of Great Britain.

1822. BERLIN. *August* 30.—Diploma of Foreign Correspondent to the Medico-Chirurgical Society of Berlin.

This Document will close the List; and is inserted for its expressive simplicity.

SOCIETAS MEDICO CHIRURGICA BEROLINENSIS.

Societas Socium Correspondentem Celeberrimum D. Jenner uno consensu elegit.

Berolini, datum die 30 Aug. 1822.

D. HUFELAND, Præses.

* See Chapter VI. p. 214 of this Volume.

The originals of nearly all these official documents were found in the possession of Dr. Jenner at his decease; and many other direct testimonials in favour of vaccination and its discoverer are contained in Reports, Addresses, and Resolutions of thanks from Medical, Civil, and Military Authorities; and the Catalogue might be greatly enlarged by the addition of commendatory letters from exalted, noble, and scientific personages, which are almost innumerable. Many of the latter have been enumerated in Mr. Pruen's Comparative Sketch.

No. II.

In this place it was my intention to have inserted a copy of the Madrid Gazette, which was issued on the return of the expedition under Balmis. As, however, a pretty full analysis of that document has been given in the text,* I am unwilling to swell this volume by printing it. The same reason induces me to withhold Dr. Sacco's paper, to which a reference has been made at p. 234. I will only state that his experiments clearly show that vaccination duly performed does not lose its influence by time; that persons who had had small-pox, as well as those who had been vaccinated, were attacked by the varioloid disease. This modified disease, when propagated by inoculation, at first produced a mild affection akin to the Variolæ Vaccinæ. A second inoculation often produced the genuine Variolæ.

In these statements I can find nothing but additional confirmation of the doctrines contained in these volumes touching the nature and influence of variolous diseases. I take this opportunity of giving another illustration of the same truth. I have in this volume mentioned the Variolous Epizootic in Bengal, and the propagation of the genuine Variolæ Vaccinæ from that source. I have since received more recent intelligence from the same quarter, which proves that more extensive inoculations from the diseased cows have produced not the mild Vaccine Vesicle, but

* See p. 78 *et seq.* vol. ii.

an eruptive disease of the true variolous character.* When the black cattle in England were affected in 1780 with a destructive variolous complaint, † there can be no doubt that inoculation from this disease would have produced similar results. Dr. Jenner at a later period found the Variolæ among the cows of a more mild and less malignant nature. He employed this mild virus, and with what success all the world knows.

I take this opportunity of expressing my regret that I have employed the word *grease* in alluding to the disease in the horse. *Variolæ Equinæ* is the proper designation. It has no necessary connexion with the grease, though the disorders frequently co-exist.‡ This circumstance at first misled Dr. Jenner, and it has caused much misapprehension and confusion.

No. III.

List of Medals struck in honour of Vaccination.

1801.—Medal presented by the Medical Officers of the Navy. Obverse, Apollo introducing a young seaman recovered from the vaccine inoculation to Britannia; who in return extends a civic crown, on which is inscribed " Jenner ;" above, " Alba nautis Stella refulsit ;" below, 1801. On the reverse, an anchor; over it, "Georgio Tertio rege; " and under it, "Spencer duce."

1803.—The Berlin Medal. Obverse, A child pointing with the forefinger of the right hand to the spot where vaccination is generally performed on the left arm. In the left hand the child holds a rose; and there is, besides, a garland of roses and

* See the Quarterly Journal of the Calcutta Medical and Physical Society, No. II. April, 1837.   † See Vol. i. p. 214.   ‡ See Vol. i. p. 242.

a cornucopia. The inscription is " Edward Jenner's beneficial discovery of the 14th of May 1796." The Reverse states the object of the Medal in words to this effect: "In remembrance of protection afforded." Presented by Dr. Bremer. Berlin, 1803.

1804.—Gold Medal of the London Medical Society, inscribed " E. Jenner. M.D. Socio suo eximio ob Vaccinationem exploratam."

The Napoleon Medal. One of the most beautiful of the series. Obverse, the head of the Emperor, with the inscription, Napoleon Empereur et Roi. Reverse, Æsculapius protecting Venus. A small cow on one side, and the implements for vaccination faintly appearing on the other, with the inscription " La Vaccine, MDCCCIV."

A Medal to promote and commemorate vaccination in the county of Sussex. It was ordered by Mr. Fuller, but I have never seen it.

1807.—A Medal struck at Bologna bearing this inscription. " Aloysio Sacco Mediol. Med. et Chir. Prof. Jennerii æmulo amici Bononienses, A. I. A. B. Ital. Reip. Cons."

Another struck at Brescia, I believe, in the same year, in honour of the same cause.

In Prussia, too, a medal, worth fifty gold ducats, was struck by order of the government, to be conferred on those who distinguished themselves in the promotion of vaccination. On one side is a head of the King with the inscription, " Fredericus Gulielmus Rex, Pater Patriæ." On the reverse is a cow with the Goddess of health, and the motto, " In te suprema salus." Round the edge, " Vaccinationis præmium.

1818.—It was likewise intended that a Medal of Jenner should be struck at Paris, to form one of that series which commemorates "*des hommes illustres de tous les Pays*."—I find a letter announcing this fact, dated Nov. 16, 1818. I cannot say whether the design has been executed. I suspect not.

INDEX

TO THE

SECOND VOLUME.

A.

C.

D.

Printed in the United States
By Bookmasters